WATER
A MATTER OF LIFE AND HEALTH

'Rahiman pani rakhiye,
Bin pani, sab soon,
Pani gaye na ubre,
Moti, manas, choon.'

Rahim the poet says, conserve water,
Without water, nothing is there,
If water goes,
Pearl, mankind, and lime will not survive.

—Abdul Rahim Khankhana,
poet and sage at the court of the
Mughal emperor, Akbar, 1556–1605.

WATER
A MATTER OF LIFE AND HEALTH
Water Supply and Sanitation in Village India

Maggie Black
with *Rupert Talbot*

OXFORD
UNIVERSITY PRESS

YMCA Library Building, Jai Singh Road, New Delhi 110 001

Oxford University Press is a department of the University of Oxford. It furthers the
University's objective of excellence in research, scholarship, and education
by publishing worldwide in

Oxford   New York

Auckland Cape Town Dar es Salaam Hong Kong Karachi Kuala Lumpur
Madrid Melbourne Mexico City Nairobi New Delhi Shanghai Taipei Toronto

With offices in

Argentina Austria Brazil Chile Czech Republic France Greece Guatemala
Hungary Italy Japan South Korea Poland Portugal Singapore Switzerland
Thailand Turkey Ukraine Vietnam

Oxford is a registered trade mark of Oxford University Press
in the UK and in certain other countries

Published in India
By Oxford University Press, New Delhi

© UNICEF

ISBN 0 19 566931 2

Typeset in Palatino in 10/12.7
by Excellent Laser Typesetters, Pitampura, Delhi 110 034
Printed in India by Roopak Printers, Delhi 110 032
and published by Manzar Khan, Oxford University Press
YMCA Library Building, Jai Singh Road, New Delhi 110 001

Contents

Maps, Figures, and Boxes

MAPS

FIGURES

All photographs courtesy Unicef

Abbreviations and Acronyms

AFARM	Action for Agricultural Renewal in Maharashtra
AFPRO	Action for Food Production
AIIHPH	All India Institute of Hygiene and Public Health
AIIMS	All India Institute of Medical Sciences
ANM	Auxilliary Nurse-midwife
bcm	billion cubic metres
BIS	Bureau of Indian Standards
CDD	Control of Diarrhoeal Diseases
CDECS	Centre for Development Communications and Studies
CGWA	Central Ground Water Authority
CGWB	Central Ground Water Board
m^3	cubic metres
COW	Contractor-oriented Work
CPHEEO	Central Public Health Environmental Engineering Organization
CRSP	Centrally sponsored Rural Sanitation Programme
CSE	Centre for Science and Environment
CSIR	Council for Scientific and Industrial Research
DFID-UK	Department for International Development United-Kingdom
DTH	down-the-hole
FAO	Food and Agriculture Organization
FC	fully covered
GDP	gross domestic product
GNP	gross national product
GOI	Government of India
GWSSB	Gujarat Water Supply and Sewerage Board
Halco	Halifax Tool Company

IAS	Indian Administrative Service
IDWSSD	International Drinking Water Supply and Sanitation Decade
IEC	Information, Education, and Communication
KAP	Knowledge, Attitudes, and Practices (study)
MNP	Minimum Needs Programme
NC	not covered
NICD	National Institute of Communicable Diseases
O&M	Operation and Maintenance
ORS	Oral Rehydration Salts
ORT	Oral Rehydration Therapy
PC	Partially covered
PHC	Primary Health Centre
PHED	Public Health Engineering Department
ppb	parts per billion
PVC	Polyvinyl Chloride
ppm	parts per million
PRI	Panchayati Raj Institution
R&D	Research and Development
RIGEP	Rajasthan Integrated Guinea Worm Eradication Programme
RWS	Rural Water Supply
SIDA	Swedish International Development Cooperation Agency
SSHE	School Sanitation and Hygiene Education
SWACH	Sanitation, Water, and Community Health (project)
TAG	Technical Advisory Group
TBS	Tarun Bharat Sangh
TSC	Total Sanitation Campaign
TWAD	Tamil Nadu Water Supply and Drainage Board
UNDP	United Nations Development Programme
VIKSAT	Vikram Sarabhai Centre for Development Interaction
VLOM	Village-level Operation and Maintenance
WATSAN	Water and Sanitation
WDS	Water Development Society
WEDC	Water Engineering Development Centre
WESA	Water Education and Social Action
WWF	World Wide Fund for Nature

Foreword

This is a timely book, chronicling in an authentic and lucid manner the ecstasy and agony that Unicef-India has experienced during its involvement in the water and sanitation programme of the country beginning from 1969. Several parts of India are currently facing a water emergency. Chennai, where I live, is struggling to find sources of water which might help to meet the minimum requirements of a population of over six million. A calamity is often not without a blessing. Tamil Nadu has become the first state in India which has made rainwater harvesting mandatory in all buildings.

The main message of this book is beautifully stated in the following words: 'If there is one lesson from Unicef's 35 years of water and sanitation programme in India, it is that without real involvement and commitment from people, including and especially women, no water or sanitation service is sustained or sustainable. As time goes on, the same will inevitably be the case with the resource itself; without the involvement of its consumers in conservation, protection, and allocation, the freshwater resource will not be sustained or sustainable either.' The chapters trace the evolution of Unicef's programmes in India, starting with the introduction of high speed waterwell drilling, designing of new handpumps adapted to local conditions, sanitation, and health issues, and ending with the ecological, social, and political threats now looming large in relation to achieving the goal of safe drinking water for all. Unicef's long experience in drinking water and environmental sanitation in partnership with central and state governments, academia, civil society and donor organizations, is summarized and analysed in this book with clarity and objectivity.

An important message reiterated throughout the book is the fact that a complex problem cannot be solved by a single or simple solution. The authors rightly point out that 'however successful a "mass" supply-driven programme appears to be, it should not be promoted without reference to the many social, economic, and environmental variables belonging to the different settings: from village to village and block to block'. This principle is illustrated with examples drawn from the work of government and civil society: well known examples like the water harvesting movements led by Anna Hazare in Ralegaon Siddhi, Maharashtra and Rajendra Singh in Alwar, Rajasthan, as well as lesser known movements like the work done in Choriya in the Tonk district of Rajasthan for reducing the fluoride content of water by use of household filtration techniques. These examples provide pointers to the way ahead. The challenge lies in making such unique examples of local-level water security systems the building blocks for an enduring national water security programme.

Maggie Black and Rupert Talbot's brilliant and perceptive analysis of Unicef's experience in *Water: A Matter of Life and Health* brings out clearly a need for a change in mindset with reference to drinking water and sanitation. I have often defined food security as 'physical, economic, social, and ecological access to a balanced diet and clean drinking water for all and for ever'. This book underlines the need for incorporating drinking water as an essential component of food security and highlights the role of non-nutritional factors like sanitation in ensuring nutrition security at the level of individuals.

To achieve a sustainable drinking water security system for India certain steps need to be followed. Water should become *Everybody's Business* (the title of a stimulating book by Anil Agarwal and Sunita Narain of the Centre for Science and Environment). This implies convergence and synergy among all ongoing efforts and programmes at the operational level. We must learn from *Dying Wisdom* (the title of another book by Anil Agarwal and Sunita Narain) in relation to water harvesting and saving technologies and equitable water-sharing methodologies. Unless there is equity in water sharing, there will be no cooperation in water saving. Social exclusion will lead to water conflicts. There is need for a decentralized approach to water harvesting and delivery systems, based on local soil and climatic conditions on the one hand, and socio-cultural factors on the other. Gender sensitivity is vital in all drinking water programmes since in nearly all parts of India, the onus of fetching water for domestic needs has been assigned to women. The quality of drinking water for dairy cattle needs monitoring since pesticide polluted water will lead to milk with pesticide residues. Drinking water programmes need to be linked synergistically with programmes relating to sanitation, hygiene, and health care. There is need for a community-centred water literacy movement which will empower children, women, and men with location-specific information on all aspects of water harvesting, conservation, quality, and sustainable and equitable use. A sustainable water security system will need appropriate blends of regulation, education, and social mobilization. Supply augmentation, demand management, and quality improvement should all receive concurrent and coordinated attention. Water should be regarded as a precious social resource and not as private property. There should be greater investment in new technologies such as bioremediation to remove arsenic and heavy metals from groundwater, solar desalination, remote sensing and water recycling methods and in the establishment of groundwater sanctuaries. (I had proposed this during the severe drought of 1979 for specially earmarked groundwater resources to be used only during emergencies.) Finally, self-destructive public policies such as free electricity for pumping groundwater, need to be abandoned as such free-for-all and unhindered exploitation of ground- and surface-water resources will only make an already difficult situation even more dangerous.

We are indebted to Maggie Black and Rupert Talbot for their contribution to the vital and urgent quest for a sustainable water future for India. This book should be read by everyone—school and college students and teachers, members of *panchayati raj* institutions (PRIs), state legislatures, and the Parliament, scientists, social workers, and activists and, above all, by policy-makers and political leaders. The book should not be merely read, but its key messages should be converted into field-level action plans. If this happens, a drinking water secure India will become a reality.

M. S. SWAMINATHAN

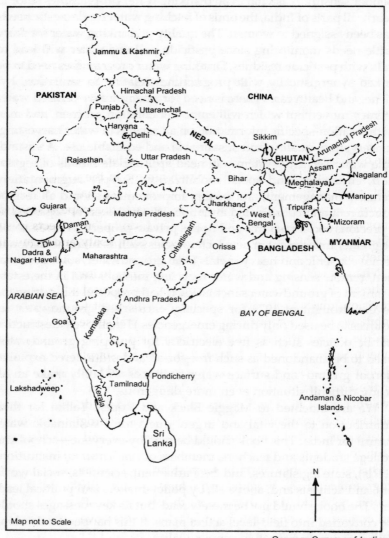

Source: Survey of India

Political Map of India

Preface

This book is the outcome of a long-standing ambition to tell the story of 'the largest rural water supply programme in the world'. Or, at least, some of that story—the part in which Unicef has been closely involved—for no single volume can do justice to the scope, scale, and innovation of India's countless water and sanitation projects in so wide a geographic spread, hydrogeological complexity, and technical variety as is to be found in this country.

When Unicef made the decision, 35 years ago, to import 125 state-of-the-art drilling rigs for a major new push in rural water supply provision, this was the largest grant, and the most technologically ambitious, that it had ever made. In a very real sense that decision was the platform from which everything Unicef has subsequently done in India was launched, not just in water supplies, but in the whole range of its cooperation. That grant caused Unicef to be seen, for the first time, as a serious development partner at the highest levels of central and state government. It opened the door to Unicef's substantive participation in many other key social programmes, such as early child development, mass immunization, and guinea worm eradication.

This is, therefore, one of the longest-running and most significant programmes in the organization's history—not only in India but around the world. It set the pace for the rest of Unicef's global water and environmental sanitation undertakings. Moreover, it served as a test-tube for all sorts of development lessons—for it has had its fair share of ups and downs. We have learnt, for example, that social advancement cannot be achieved without technical excellence, but that the converse is also true. The best technical input is no guarantee of improved social well-being unless the community, especially its womenfolk, have been mobilized and expressed commitment. We have learnt that providing safe water does not necessarily mean that safe water will be used; that people do not realize how hygiene reduces disease unless it is properly explained. We, together with our partners, have had to learn to be far more balanced in our approaches, between technical, social, economic, cultural, environmental, and even ideological considerations.

Unicef's support for water and sanitation has always been aimed at improving children's and women's health. But there have been many disappointments and false starts in the search for that aim to be fulfilled. All development is a quest, and this programme may be seen as a casebook example of the evolution in thinking and practice that has to occur. The story of village water supply and sanitation in India also supports the argument that true partners in the process have to be committed for the long haul, with all its twists and turns.

Of such partners, there have been many. Pride of place naturally goes to the Government of India (GOI), whose commitment to rural drinking water supplies in successive Five-Year Plans has provided the framework for this particular quest. But others have also played vital roles. From the Church of Scotland missionaries who first experimented in India with modern drilling rigs and handpumps to Unicef's key bilateral donors, notably Sweden, Switzerland, Denmark, the Netherlands, and the United Kingdom (UK); from the State Public Health Engineering and Water Boards and district officials who have thrown their efforts behind the implementation to the manufacturers of pumps and rigs and their quality inspectors. From the large numbers of local non-government organizations (NGOs) who today carry out mobilization for water and environmental sanitation on the ground to the many committed individuals in all the different institutions and organizations who have contributed so much to the programme. They all deserve our profound congratulations.

Today, in spite of all the years, the investments, and the approaches, Indian villages still face severe drinking water problems. In an ironic twist of fate, the take-off in hard-rock drilling which Unicef helped foster is leading, as we all know, to serious groundwater depletion. This also has alarming outcomes for water quality: millions of cases of fluorosis have emerged, and arsenic threatens groundwater in

West Bengal and elsewhere. Sanitation facilities have yet to pick up in most parts of the country. There is also a question mark over the sustainability of what are now vast numbers of local water supply installations: how are communities to take over service delivery responsibilities and carry them out effectively?

The long haul is far from over. In fact, as this book shows, the solution of one set of problems has an awkward way of throwing up a whole new generation of problems of many different kinds. As the pressure mounts on India's freshwater resources, this is a timely moment to take stock. This Unicef-sponsored book has not been produced in a spirit of self-satisfaction. Because this has been for Unicef a very special programme, we have decided to take a long, hard look at the record. There is a great deal in these pages of which we and our partners can feel justly proud. And there is much from which to learn—on the basis of which the great Indian adventure in rural water and sanitation can, and will, go forward.

Maria Calivis
Representative
Unicef, New Delhi
India

February 2004

Acknowledgements

This book, a collaboration between a long-time Unicef water practitioner and the author of many books about Unicef's work, was first mooted in 1998. At the time, a major evaluation of Unicef-India's water and environmental sanitation programme since its inception in 1966 was underway, with support from certain donors: the Department for International Development (DFID, UK), the Netherlands Ministry of Foreign Affairs, and the Swedish International Development Cooperation Agency (SIDA).

The then Unicef-India Country Representative, Alan Court, agreed that the remarkable story of the organization's role in the most extensive rural water supply programme in the world deserved to be told in a more accessible form. Circumstances dictated that work could not begin on the draft until four years later, by which time the crisis overtaking water supply in village India was growing ever more acute. Thus the delay has favoured a broader analysis, extending beyond areas of policy and practice in which Unicef itself plays an influential role. Our first acknowledgement, then, is to Alan Court, and to his successor as Unicef Representatives to India, Maria Calivis and Cecilio Adorna who have given the project unstinting support.

Although this book was commissioned by Unicef, the authors bear all responsibility for the views expressed. We deeply appreciate the organization's willingness to respect our independence of mind. Unicef recognized that a self-congratulatory account would lack credibility and prevent lessons learned from being properly brought out. This book is, therefore, an attempt to view the Unicef-related experience of water and sanitation programmes in village India over 35 years as coolly and objectively as possible, distinguishing between what has worked and what has not. If some of the conclusions are not shared by certain readers who themselves played a part in the events, that is in the nature of an account of this kind. We hope that such readers will appreciate that no criticism is intended of any individual or institution, and will accept differences of view in the spirit of enquiry which is the only way in which sound development policy and practice can emerge.

The programme had one important unanticipated outcome. The high-speed waterwell drilling technology Unicef imported on humanitarian grounds in 1967, and later helped to transfer into the Indian technical and commercial environment, has itself contributed to today's water scarcity and quality crisis. Although the use of down-the-hole-hammer drilling in drought-prone villages soon became controversial among some NGOs, at the time it was widely seen as a brave and noble enterprise. Many of those involved in decisions surrounding the transfer of this technology, inside and outside Unicef, can claim to be godparents—for good or ill—of the whole rural water supply venture. Their role, as pioneers, mentors, colleagues, and friends needs to be acknowledged here.

In particular, we would like to mention just a few of those who belonged to the heroic tradition of missionary endeavour in India, then in its declining years: Clement Ferer, Charles Heineman, John McLeod, Clement Moss, Gifford Towle, Chris Wigglesworth, and Peter Wood. Among those Indian colleagues who were active from the earliest stages of the programme and who have stayed with it in their many ways, we would like to pay tribute to Mansoor Ali, Raj Kumar Daw, Dilip Fouzder, Vishwas Joshi, J. N. Kathuria, and Sanjit (Bunker) Roy. At the international level certain other Unicef colleagues also played a part: M. Akhter, Gordon Alexander, Martin Beyer, Gordon Carter, Colin Davis, Glan Davies, Nigel Fisher, Ken Forman, Kul Gautam, Ken Gray, John Grun, David Haxton, E. J. R. Heyward, Raymond Janssens, Greg Keast, Malcolm Kennedy, Ken Mcleod, Ashok Nigam, Henk van Norden, Jon Rohde, John Skoda, Philip Wan, and Eimi Watanabe. This is by no means an exhaustive list, but a personal tribute to those whose ideas and example have influenced the mainspring of the story told in these pages.

During the course of our own, separate, accumulating experience of Unicef and of India, a number of other individuals have shaped our thinking in ways that find direct or indirect expression here. These include: Michael Acheson, Vinod Alkari, Anil Agarwal, Saul Arlosoroff, Joe Aspdin, Erich Baumann, Paul Calvert, Oscar Carlson, John Chilton, William Cousins, Palat Mohan Das, M. M. Datta, Mahesh Desai, Anu Dixit, Gourishankar Ghosh, Arden Godshall, R. Gopalakrishnan, Biksham Gujja, Suranjan Gupta, Jim Howard, Leela Iyengar, Ramesh Jagtiani, Tim Journey, Krishan Kalra, T. Kanagarajan, T. S. Kannan, Ashok Khosla, Peter Kolsky, A. P. Mangalam, Frank Morrison, Arun Mudgal, Sunita Narain, Ramesh Panda, Ishwarbhai Patel, Bindeshwar Pathak, Medha Patkar, N. S. Prasad, Lalit Raichur, B. B. Rau, Raymond Rowles, B. B. Samanta, N. C. Saxena, A. K. Sengupta, P. K. Sivanandan, J. E. Sokkiah, John Staley, A. N. Susheela, Gordon Tamm, and Vincent Uhl.

In the course of field visits, undertaken specifically for this book, to West Bengal, Rajasthan, and Karnataka, we would like to acknowledge the help of: Carrie Auer, Chandi Dey, Satish Kumar, Manoj Kumar, N. S. Moorthy, Chandan Sengupta, S. N. Singh, William Thompson, and Dev Vaish, as well as many officials at district and block level and NGO personnel too numerous to mention. Last, but by no means the least, are the *sarpanches* (heads of the elected councils), village and school leaders, members of sanitation committees, handpump operatives, community health workers, and ordinary men, women, and children who shared their experiences and perspectives—and their wisdom.

Within Unicef's India Office, the following people helped review the text and bring the project to fruition: Erma Manoncourt, Susan Bissell, Sumita Ganguly, Lene Hansen, and Savita Naqvi; special thanks in this context also go to Shiva Kumar. We would also like to express our gratitude to M. S. Swaminathan, eminent environmentalist and father of the Green Revolution, for contributing a foreword which lends distinction to our work. Finally, we would like to thank Sarla Varma and Parminder Singh for their administrative assistance, Alka Malhotra for contributing to the research, and the editors at the Oxford University Press, India for their enthusiastic support throughout the preparation of the book.

MAGGIE BLACK   RUPERT TALBOT
Oxford, UK   New Delhi, India

1

Water: A Matter of
Life and Health

Life, prosperity, and civilization revolve around water in the Indian
sub-continent. Children and lovers sing of the fortune and happiness
water confers, the devout celebrate water's cleansing and health-
giving properties in their rituals and prayers. But this water, so
central to everything material and spiritual that happens in India, is
a feckless and ephemeral affair. To be sure, great rivers flow majes-
tically through the landscape, and their perpetual renewal of life,
as well as their absorption of death, earn them deep veneration. But
the behaviour of India's waterbodies—from the mighty Ganga to the
smallest stream—is fickle. Everything depends on the rains.

When the monsoon clouds release their deluge, dried-up riverbeds
begin to trickle, streams swell, tanks, wells, ponds, and lakes fill up,
and everywhere there is the splash and hum of water overflowing.
Landscapes that were stark in their brown aridity spring into life,
green first dusting the tops of the furrows and the slopes of the hills,
and then crops, saplings, grass, flowers, and vegetation bursting
out. The monsoon is depicted in Indian literature as the season of
love, when plants and people all receive a new lease of life. But the
monsoon, however predictable a meteorological phenomenon, is
notoriously temperamental.

The rain pours down tempestuously in some areas, bursting banks,
submerging crops, drowning homes and livestock like a goddess of
destruction. In West Bengal and Orissa water is a part of the landscape.

There are puddles the size of small seas lapping everything around them, conferring an unparalleled fertility. In other areas, the promised renewal of life is an illusion. In Gujarat and Rajasthan soil crumbles into dust as clouds pass overhead towards the horizon without disgorging a single drop. For many farming families failure of the rains is a calamity too painful to contemplate. 'There is nothing to do but hope for rain,' says a Rajasthani farmer when asked how his family will cope with a fifth successive year of drought. Without rain the tanks will run dry, and so will the wells. There will be no water for irrigation, drinking or domestic purpose, no fodder for the cattle, goats or camels and, eventually, when the store is empty, there will be no food. No one wants to face that prospect for themselves or their children.

The perennial sense of urgency surrounding the rains in India, especially for its millions of small farmers and livestock herders, is hard to communicate to those who take for granted that water is an abundant resource available at the turn of a tap. With 20 per cent of the world's people, India is forced to meet all of their needs for water .rom only 5 per cent of the world's available supply (Glieck 1998). This makes water a supremely precious resource in India, and how it should be managed and distributed has always been critical to the organization of human affairs. In recent decades, issues surrounding water have become increasingly contentious, as claims on finite supplies in a shrinking national water pot grow more voluminous and insistent by the day. The strain is felt throughout India's towns and cities, on its institutions and infrastructural capacity, and in its burgeoning industries. But nowhere is it more acute than in those parts of rural India prone to drought, where, if streams and wells dry up, life itself is threatened.

The modern story of improved rural drinking water supplies in India began in the late 1960s. It grew out of an emergency, and the humanitarian response to the plight of its victims. Crises such as droughts, famines, earthquakes, floods, calamities, and confrontations over the use of power and resources within society, often act as a catalyst for innovations in policy and technical change. In India, water occupies such a dominating place in the panorama of life that, down the centuries, it has provided an arena in which epic dramas are played out. The drought of 1966–7 was the last occasion in which a famine—in Bihar—was declared by the authorities. The drought exposed important vulnerabilities: in food production for a growing population dependent on the vagaries of rainfall to eat; and in the failure of water sources which, under extreme environmental pressure, dried up or sank beyond reach. The emergency programme mounted at that time led, in due course, to a major expansion of the national rural drinking water supply programme in which Unicef became the key external partner. Within a generation, the programme had revolutionized access to water in the hard rock areas of the country and brought supplies from deep in the earth to millions of households.

Yet the programme has not been able to banish the spectre of water shortage in rural India, and drought continues to beset many parts of the country when the monsoon fails. Those wrestling to bring the fertility of the earth into a balance with its annual rejuvenation by untameable natural forces can never regard their contest as definitively won. Nonetheless, there have been important gains.

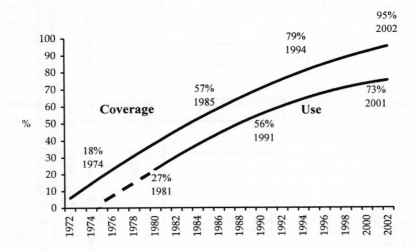

FIGURE 1.1: Rural water supply coverage and use

Sources: International Drinking Water and Sanitation Decade—Reviews of National Progress, by WHO, UNICEF; 10th Plan document. Census of India, National Family Health survey (NFHS), Multi-indicator Cluster Survey (MICS).

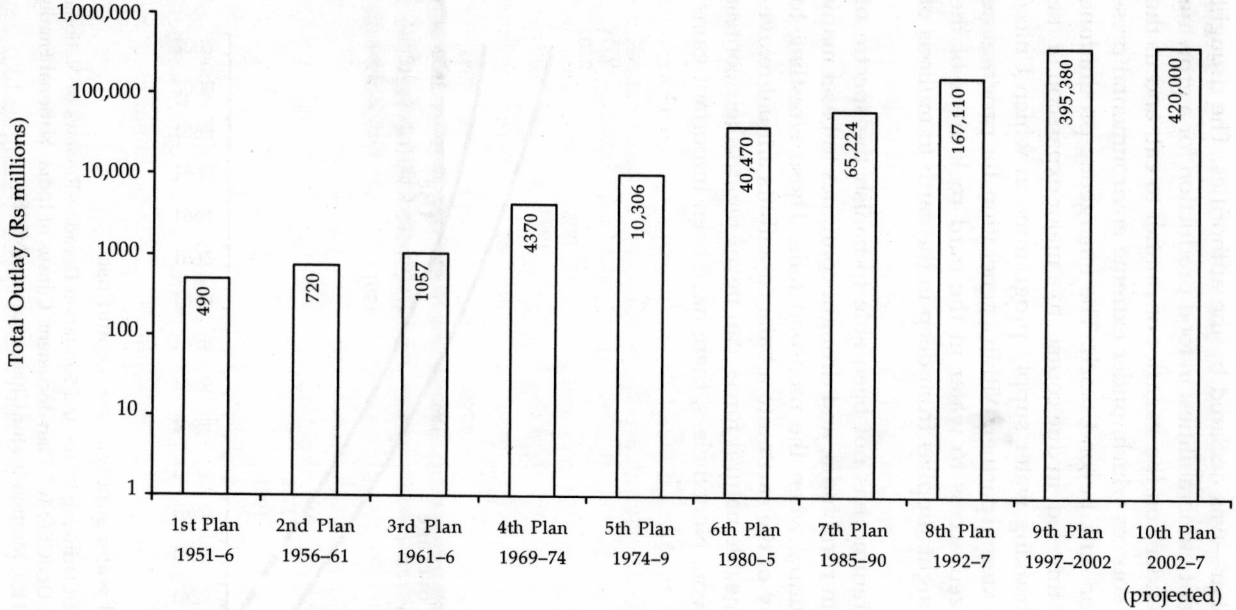

FIGURE 1.2: Government investments in water supply and sanitation

Sources: Planning Commission, Govt. of India.

Note: 1. Outlays shown are central plus state investments at current prices.
     2. The scale of the Y Axis, Total Outlay, is logarithmic, not linear.

By the end of the twentieth century, 'coverage' of safe water supply to India's rural people had grown dramatically: from 18 per cent in 1974 to 95 per cent in 2002 (GOI 2002). This came about because of major governmental investments. The countrywide programme has been a boon for rural families—especially for women and children. Its contribution to improved public health in the countryside, although difficult to quantify, is beyond doubt. Over large parts of the peninsular slab which covers central and southern India, handpumps with a peculiar slanting box at their head nowadays have an ubiquitous presence, along roadsides, in villages, and in the more modest parts of the towns. Over 3 million of these distinctive 'India Mark II' pumps have now been installed. Clusters of brightly-dressed women and girls with brass and plastic pots queuing on concrete aprons for a turn at the handle is as common a sight as tea shops and marigold garlands. This is something over which all those involved in the development of the handpump–borehole for rural water supply can feel justly proud.

However, like the behaviour of water, the course of social change for which these installations are the visible symbol has proved unpredictable. The smoothly flowing stream of improvement represented by the impressive water supplies statistics has its eddies and backwaters, and its hidden obstacles below the surface. India is dependent for her freshwater needs on groundwater, grand schemes to shift large volumes of surface water between river basins notwithstanding. Groundwater provides 80–90 per cent of domestic water supply in rural areas, even though the proportion drawn for this purpose is small: less than 5 per cent of extractions. It also provides for 50 per cent of urban and industrial demand for water (Nigam et al. 1998). The successful exploitation of groundwater—a source of water regarded as 'safe' because it is less vulnerable to contamination than surface water running in streams and collecting dirt and germs along the way—has been the key to drinking water provision in modern India. This is especially the case in hard rock areas where water is scarce and difficult to reach, and where predominant surface streams are seasonal.

There is, however, a less rosy side to the picture. The extent of water-well drilling in the past 25 years has helped precipitate a gathering water crisis. The advent of high-speed drilling technology in the early 1970s ushered in an era of extensive groundwater extraction for irrigation. Around 50 per cent of the country's irrigated area is now fed by groundwater, rather than by surface water from canal networks, tanks, and reservoirs. Over this period there has been a huge increase in power-driven pumping from wells: from around 25,000 at mid-twentieth century, the number of energized wells in India has now risen to over 20 million (Viksat 2003). Where diesel used to be more common, most of these wells are now powered by the cheap energy provided courtesy of rural electrification: the electric pump is now an item within the consumer reach of millions of Indian farmers. The explosive demand for groundwater has also been fed by the need to grow food for an ever larger population, and by an increasing focus on water-thirsty cash crops as a means of raising agricultural incomes.

The exploitation of the resource has taken place at a speed which does not allow time for the water table to recoup its losses. Its inevitable accompaniment has, therefore, been the drying up of shallower aquifers, a reduction of water flows in rivers and streams, and the progressive deepening of wells. Where a depth of 10 metres used to suffice for a plentiful supply, a depth of 80 metres may now be required, and the rate at which the water table is declining has accelerated markedly in recent years (UNICEF 1998). With the poor monsoon of 2002, state departments for rural development—in Karnataka, Andhra Pradesh, and Maharashtra, for example—were desperate to increase the output from existing wells rather than undertake yet more indiscriminate drilling. In October 2002, all through the landscape of drought-stricken Rajasthan, crude derricks and piles of stone indicated attempts to deepen dug and blasted wells still further. This is a traditional response to wells running dry under climatic duress, but the response is nowadays itself helping to deepen future crisis.

The annual monsoon not only fills rivers and streams, but as rainfall seeps into the soil, recharges underground aquifers. The problem today is that this recharge process is inadequate in comparison to the rate of extraction, especially in the one-third of the country—99 districts in 13 states—classified as drought-prone (Nigam et al. 1998). The uptake of groundwater has, in many of these places, crossed the limit imposed by natural rates of renewal. Around 550

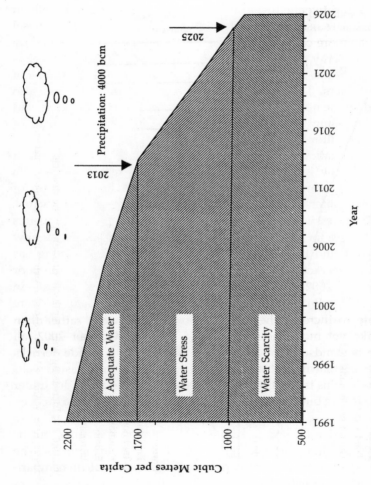

FIGURE 1.3: Annual per capita freshwater availability
Source: After Falkenmark (Unicef time line, 1998)

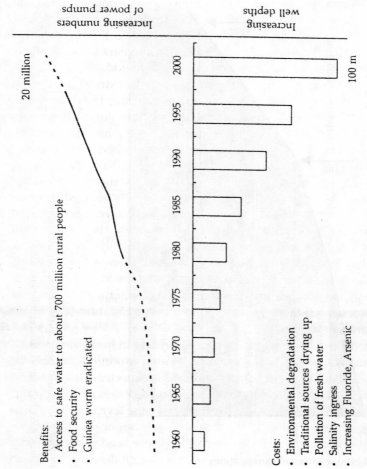

Benefits:

- Access to safe water to about 700 million rural people
- Food security
- Guinea worm eradicated

Costs:

- Environmental degradation
- Traditional sources drying up
- Pollution of fresh water
- Salinity ingress
- Increasing Fluoride, Arsenic

FIGURE 1.4: Increasing well depths in drought-prone districts.

Source: After VIKSAT—Vikram Sarabhai Centre for Development Interaction.

blocks[1] in Andhra Pradesh, Gujarat, Haryana, Karnataka, Madhya Pradesh, Maharashtra, Punjab, Rajasthan, Tamil Nadu, Uttar Pradesh, and West Bengal have been classified as 'dark' or critical in relation to the prospective exhaustion of finite supplies of groundwater within the coming years. A further 600 have been designated 'grey' or semi-critical (Nigam et al. 1998). In some of these areas—notably in Gujarat and Madhya Pradesh—the problem is exacerbated by the mining of fossil water located several hundred metres below the surface. This kind of source, which has taken thousands of years to accumulate, cannot be recharged: once used it is gone for good.

The pressures on the supply are leading to extreme measures. Water trains regularly ply Indian railways during the dry season, shifting water from less drought-prone areas to those such as Rajasthan, Gujarat, and Tamil Nadu where shortage is endemic. To resort to conveying water by vehicle instead of by gravity-fed pipeline or by deepening wells is normally an emergency relief measure, but in some places—predominantly in Maharashtra—it has become a regular feature of water provision for the most severely affected communities. In some 'dark zone' areas villages are obliged to depend on the refilling of local wells by water tankers. Since the tankers only make their rounds every several days, the water these people drink is rarely 'fresh' in the usual sense of the word. However, at least this kind of supply is not likely to be chemically tainted. In many locations, depleted aquifers are showing high concentrations of contamination from fluoride, arsenic, iron and nitrates and, in coastal areas, from saltwater intrusion.

Thus, the triumphant success of well-drilling in India is now helping to shape a potential catastrophe. Within around 10 years, the annual availability of freshwater per head is expected to drop below 1700 cubic metres, the internationally agreed measure of 'water stress'; by 2025, to below 1000 cubic metres—the level used to define 'water scarcity' (UNICEF 1998). This will be brought about by population growth and rising water consumption per head, independently of how the wilful monsoon behaves. Even though the principal cause is the thirst of crops, people—especially the poorest people—will

---

[1] 'Blocks' here include mandals, talukas, and, in Maharashtra, watersheds, as identified by the Government of India (GOI) Central Ground Water Board.

suffer the worst. If their granaries and wells are empty, they cannot resort to the market to buy their way out of trouble: they do not have the means. These people, too, will bear the brunt of the crippling diseases which are increasingly emerging in pockets of the country, from Rajasthan in the west, to Punjab in the north, from Assam in the east, to Tamil Nadu in the south (Jamwal and Manisha 2003). Around 62 million people are estimated to risk illness from drinking too much fluoride in their water. In West Bengal, 13 million people face the possibility of arsenic poisoning. Neither of these contaminants can be detected easily by taste or appearance. Iron and salt in water flowing from thousands of other public handpumps and standpipes do not pose an equivalent threat to health. But they make the water taste unpleasant or spoil laundry, so people prefer to use other sources—dug wells or streams—whose risk of bacteriological pollution is higher.

The main topic of this inquiry into water supplies for village India is not the role of water in India's overall socio-economic development, but a subsidiary, albeit important, component of this very much larger subject. It is principally concerned with the interactions between the supply of rural drinking water and sanitation services, and the health and well-being of countless underprivileged children and families. This is the area to which Unicef, in partnership with government and non-governmental organizations (NGOs), has contributed, and it is through the lens of Unicef's 35 years of cooperation in this area that the story is principally told.

However, one part of the Indian water picture cannot be addressed without reference to others. Drinking and washing are only two of the uses for which any family needs water, and no reasonable quality of life is assured by exclusive attention to minimal domestic needs. Many of today's water experts insist that water is indivisible and its uses and functions should be considered in tandem, not tackled separately under different headings by different sets of authorities. They argue that unless water policies for all uses and user-groups are integrated, this finite and precious resource cannot be efficiently managed or equitably shared. Instead, there will be constant competition over water, between farming families and urban dwellers, environmental conservationists and industrialists, minorities living off natural resources and entrepreneurs seeking to commodify the resource base for commercial gain. And it is undoubtedly true that, in India, water resources are under pressure from so many different directions that unless they are carefully managed and different claims upon them fairly adjudicated, water-related disputes will proliferate and become increasingly bitter in the coming years.

The threat of water scarcity to future Indian prosperity, health, and environmental integrity, provides the over-arching context for this story. The part Unicef has played, and continues to play, in this ongoing drama in the context of its concern for children, women, and public health, is necessarily small. Nonetheless, it provides a window on the much larger picture from the vantage point of a deeply involved organizational participant, which is, at the same time, both an outsider and an insider. Some of the outstanding questions are uncomfortable. Why is it, that after 35 years, hundreds of thousands of villages are still without a safe, dependable, year-round supply of water? Why is the squandering of groundwater and the downward plunge of water tables continuing unchecked, putting at risk all the gains in water-related health and well-being? Many of the lessons from recent decades have yet to be fully understood and absorbed.

The Indian government has repeatedly committed itself to the ultimate provision of 'safe water and sanitation for all'. Most recently, it has signed up to the UN Millennium Development Goal of reducing by half the numbers of those without safe water and sanitation facilities by the year 2015. An extraordinary effort will be needed to realize these objectives, and at the same time enable its poorest rural citizens to retain their hold on a share of a vital and shrinking resource. In that effort, Unicef will continue to play its part. And it is the poorest families and the poorest children whose rights Unicef primarily exists to defend.

From time immemorial, India's villages have struggled to be self-sufficient in food and water. Given the pattern of India's rainfall— the monsoon's torrential downpours providing a brief season of fruitfulness followed by long, dry months when the harvest must be stretched—this struggle has been intense.

In different parts of the sub-continent, it has taken different forms. Every conceivable topographical and hydrogeological variation is present in this region—from the high Himalaya mountains to the coastal swamps of Kerala, Orissa, and the Ganga delta, from the well-watered alluvial plains of the Punjab to the deserts of Kachchh and Rajasthan. Climatically, the range runs from extremely hot to cold, and extremely wet to dry, with high, cold deserts as well as zones of low-lying tropical humidity as uncomfortable as any in the world. The country also contains every conceivable diversity in rural settlement pattern, lifestyle, and socio-economic status. These include scattered hill-side village habitations, lush tropical waterlands with dense clusters of houses nestling beneath the palm trees, fortress-like desert communities of crenulated stone and mud, forested tribal areas where cash rarely changes hands, well-heeled mini-towns surrounded by fertile alluvial soil, whose farmers trade in distant markets, sit in spotless linen on their *pukka* verandas, and park their tractors beneath the trees.

As well as being vast and infinitely diverse, India is a very old country. The Harappan civilization of 3000–1500 BC in the Indus Valley is one of the very earliest in the world to build towns and trade internationally. Without sophisticated water management techniques, this would have been impossible. Wells and irrigation works on the Indus can be dated back to 2600 BC, and archaeologists have found evidence of the use of lift and surface canals for irrigating winter crops (Agarwal and Narain 1997). Even in such distant times, people realized that if water descended from the skies only in violent bursts during a brief period of the year, ways must be found to capture and control it so that its life-giving and food-growing benefits may be spread over the dry season.

By the time of the reign of northern India's first emperor, Chandragupta Maurya (321–297 BC), people were thoroughly conversant with rainfall regimes, soil types, and water management in different terrains. This is attested by the *Arthasastra*, the treatise on all aspects of government administration, attributed to Chandragupta's steely and all-powerful administrator, Kautilya. The *Arthasastra* produced a system of ecological classification of terrains in Chandragupta's domain—forests, village areas, mountainous areas, wet or humid areas, drylands, plains, and uneven lands—which covers most of the span of ecological environments embraced by the modern Indian state. Rivers, lakes, and springs were harnessed by embankments, government superintendents were instructed to build tanks and wells in order to increase agricultural output and, thereby, state revenues. Kautilya was aware that the most practical means of obtaining water in waterless (*anudaka*) areas was to dig wells with underground springs as their feeders (Agarwal and Narain 1997). In regions blessed with a good supply of water, notably the Indo-Gangetic plain, agriculture could be expanded by irrigation to put land under different uses: flower and vegetable gardens, fruit orchards, wet crop fields, and the cultivation of edible roots.

As the centuries progressed, India continued to develop hydraulic technologies appropriate to different terrains, rainfall and surface water flow patterns, farming and social requirements. Dams, bunds, and embankments were used to divert streams into the fields; some were very substantial structures, creating sizeable lakes irrigating thousands of hectares. In arid areas, they also channelled water into cisterns and storage tanks dug deep underground to prevent evaporation. In some areas, wells were dug, sometimes to depths of over 100 metres. Some of the ancient step-wells in Rajasthan, which allowed women to descend with their water pots far down into the ground, are among India's most striking architectural monuments. Elaborate lever and pulley mechanisms were invented for lift irrigation, some of which can still be seen in action. In flood-plain areas around the Bay of Bengal, inundation channels were used to control and divert floodwaters and silt for agricultural use. In the hilly North-East, bamboo pipelines carried water from springs over difficult terrain to drinking points and plantations. This almost bewildering array of ecologically adaptive water husbandry techniques gives India the right to claim, historically, world leadership stature in water conservation and management.

While communities in these different environments applied tenacity and ingenuity to solving their water problems, on a larger scale water management was manifestly an instrument of political and economic power. From Chandragupta onwards, India's rulers, whether emperors, maharajahs, or British colonial masters, all set up as hydraulic mughals. They dammed lakes, built tanks, and used engineering prowess to conserve water during the wet season for cultivation in the dry. Early in the twentieth century came the first signs of the coming clash between the state pursuit of massive technological manipulation of water, and the attempts of communities to sustain their traditional, smaller-scale water systems in the face of disregard—even outright destruction in cities such as Chennai (then Madras)—by the British authorities (Meller 1990). The vast complex of canals the British built in the Indus Valley transformed agriculture in the Punjab, but was to have many unanticipated outcomes—not least the subsequent competition over Indus waters between India and Pakistan. Waterlogging and salinization of soils has dogged many major irrigation works, demonstrating that the technological manipulation of the landscape exacts an ecological and human price which is impossible to predict with certainty, and that it invariably falls on those least able to pay. Misgivings about large dams and waterworks surfaced long before the end of the British era: Rabindranath Tagore wrote a play in the 1930s in which he symbolized colonial rule through the dam, and Gandhi's struggle for freedom from dependency and control as the liberation of the river (Ward 1997). However, the colonial authorities' administrative inheritors, the irrigation aristocrats of independent India, not only followed the preset path, but followed it more energetically (Agarwal and Narain 1997). The construction bonanza which began in the first half of the twentieth century accelerated rapidly in the post-Independence period.

India's first independent administration, that of Prime Minister Jawaharlal Nehru, embarked on a major effort to modernize India by means of state intervention. The path he chose was driven by socialist ideology, and envisaged a process of rapid industrialization to expand national economic growth. Nehru's view of the part water should play in his drive for industrial progress was epitomized by his well-known attitude towards the might of modern hydraulic engineering. In 1954, he said of the Bhakra–Nangal dam towering over a Himalayan gorge: 'As I walked round the site I thought that these days the biggest temple and mosque and *gurdwara* is the place where man works for the good of mankind. Which place can be greater than this?' (Khilnani 1997). He epitomized in his philosophical determination to uplift the common man the conviction that both the grand and the humble schemes, like the proverbial lion and the lamb, could lie down together. At the time, his view that dams would be the 'modern temples of India' reflected a widely shared belief in the transformational capacity of modern technology and large-scale infrastructure. Backward rural classes were not seen as having any role in 'development'. This was governed by the overlords of modernization, with their huge templates for land and water manipulation and their multi-million dollar projects. The poor would become the beneficiaries and grateful recipients of 'development' as the fruits of economic progress proliferated and became available for redistribution, via programmes of mass relief and subsidized service delivery.

The tide of modernization had many blessings of scientific and technological advance; but there were also losses of experiential knowledge and wisdom. Local systems for water management plunged into decline, overwhelmed, on the one hand, by India's love affair with concrete and engineering prowess and, on the other, by the notion that people at the bottom of the rural heap were ignorant and backward and should be provided for by hand-outs from the state. Some people and organizations rowed against this tide. Most of them belonged to the Gandhian tradition of advocating small-scale local solutions not only to water service provision but also to other development problems. India has a very long tradition of voluntary association between people bound by location, caste, religion, or any other kind of mutual interest (Jaitley and Daw 1995). Alongside some of the Christian missions and the groups inspired by Gandhi and his spiritual precursors and disciples, a patchwork of movements within civil society began to take up a developmental, anti-poverty, social justice, mantle. Their perspective began from where ordinary people were at and what they needed in the way of solutions to their problems. They were not driven by macro-considerations of gross national product (GNP) per capita, industrial growth, spread of modern communications and energy, or national food-grain output. Their inheritors are those community groups and voluntary bodies which are sincerely committed to building development not for, but with the people in ways that would allow the latter to profit from 'the new' without having to sacrifice those methods which had served them well in the past.

Development in India has always been a contest—sometimes bitter, sometimes benign—between these two schools of thought. But the parameters of this ideological stand-off were not as clear in the heady days of newly-independent India as they were to become later. The battle of ideas is mirrored by the history of the relationship between the government and the NGOs, which oscillates from hostility and suspicion to beneficent cooperation and mutual respect. In the interstices of programmes and planning documents, these two developmental approaches—crudely, the 'top-down' and the 'bottom-up'—overlap. Water management is no exception. In fact, the story of water resources management in independent India can be seen as symbolic of this ongoing debate about development theory and practice. Policy-makers, officials, dedicated professionals, NGOs, and activists are still struggling today to fuse these two traditions and marry their best elements together.

By the middle of the twentieth century, population growth was putting pressure on the environment, demanding more food, and more water. There was growing anxiety about water scarcity, and continuing dependence on massive importation of PL480 US surplus food stocks (8 million tonnes of wheat in the drought year of 1966) (Black 1987). Most villages and farmers maintained their tanks, wells, lakes, springs, and rivers, irrigated some of their crops by age-old methods, conserved enough drinking water and food, and, except in major crisis years, managed to pull through. But the big picture was increasingly worrying the authorities: the lack of national food security, the apparent inability of the country to grow enough to feed itself in the mathematical sense of volume of grain divided by numbers of people, the belief that semi-subsistence farmers were grossly inefficient in their exploitation of land and water resources. This naturally drove them towards the Green Revolution agricultural technology.

The new 'miracle' hybrid strains of wheat and rice were to provide the harvests which would make India a food surplus, rather than a food deficit, country. But these seeds imposed new burdens on the environment: they had a rapacity for irrigation and other capital-intensive inputs—fertilizers and pesticides—to guarantee abundance. Their use required the construction of large hydraulic projects which, whatever their benefits, also had unintended consequences, altering flows and degrading freshwater in lakes and rivers. By the late 1960s new strains on water resources were already starting to tell, both at the national and at the more local level of farming families and communities.

The 1966–7 famine in Bihar was a landmark in the annals of Indian development, a time of national spasm and post-Independence angst. The country had proved unable to feed its own citizens following a natural disaster, and been humiliatingly portrayed to the outside world as a beggar dependent on international charity. This led to profound national self-examination, and a determination that 'never again'. George Verghese, Prime Minister Indira Gandhi's Information Advisor, wrote: 'The famine has been a revelation, a trial, a shame; but also an opportunity and an awakening. It has transformed some of the inertia of the past into energy' (Verghese 1967). In 1968 there was a bumper harvest, the first in which the new hybrid seeds helped to fill the national granary. Within a few years, India became a food-surplus instead of a food-deficit country. India has never since declared a famine or suffered a major food shortage disaster, nor faced drought without sufficient grain in hand to avoid imports of food from elsewhere. Its capacity to handle disasters with the potential for human and livestock catastrophe on a large scale have, since then, become one of its proudest development achievements. The importance of this accomplishment has helped to mask the social and environmental price that has been paid in less visible areas of the rural economy.

Quite separately from agricultural and food policy concerns, an expanded national drinking water supply programme was also set in motion, under the Central Public Health Environmental Engineering Organization (CPHEEO) in the Ministry of Health. There had earlier been a national water supply and sanitation programme, launched in 1954, to build village systems requiring a higher level of technical skill in design and construction than traditional dug wells; but the resources allocated were modest (Jaitley and Daw 1995). After the 1967 emergency, priority to village water supply was enhanced. At the end of the line, people who need water to drink and people who need water for irrigation are the same people and had previously used the same sources for both purposes. But the different uses—for survival and health on the one hand and for agriculture on the other—were now compartmentalized in central and state policy as if they were distinct. This was a turning point: the moment at which programmes and policies relating to water in its different uses diverged and were sent off upon different tracks. That a tension might develop between agricultural policies relying on heavy rates of water consumption, and the need to secure the dwindling water supplies of modest rural consumers and small-scale irrigators, was not appreciated at the time.

An arbitrary division between 'water for agriculture' and 'water for health' is more conceptually valid in a temperate climate where cultivation is rain-fed and the demand for water services is confined to domestic, industrial, and recreational needs. In Europe, for example, water supply is not a livelihood issue for ordinary citizens. In a country with a monsoon climate, where water has to be stored in rivers, reservoirs, tanks, cisterns or in the ground for all dry-season livelihood and domestic needs, the picture is entirely different. Access to water resources and water services for ordinary rural citizens is not just a matter of life and health, but a matter of economic survival—the difference between relative prosperity and ruin. Nor was the finite nature of the resource respected when the policies went off on different tracks. Water was still seen as an abundant and free natural asset even if it was becoming more difficult to access. No consideration was given to the need for a set of mechanisms whereby the requirements of different uses and users could be efficiently and democratically moderated.

At the time, no warning bells which might have influenced key policy-makers were sounded. Concern about environmental integrity and the need to conserve natural resources for future generations was still in its infancy internationally, and had yet to appear in a home-grown identity on India's domestic agenda. In the 1970s, when serious official commitment to rural drinking water services began, the international climate fully favoured the compartmentalized approach. The impetus for improved or 'safe' domestic water supplies was intellectually driven by the perceived connection between inadequate or 'unsafe' supplies and epidemic disease. This idea had been ingrained in the public health engineering mind ever since the sanitary revolutions of the nineteenth century banished cholera in the West by installing drains, piped water supplies, and sewers in the pestilential towns and cities. The social costs of 'unsafe' supplies—high levels of morbidity from drinking contaminated water, epidemics of life-threatening diarrhoeal disease spreading from the slum quarters to the smarter parts of the town—provided the justification for drinking water supply investments: hence, the authority over the Indian programme of the Ministry of Health. The idea that supplies must, above all, be 'safe'—another hangover from the experiences of the industrialized world—meant that the almost exclusive focus of rural drinking water programmes was on the exploitation of groundwater by drilled handpump–boreholes. Open sources such as wells, tanks, and ponds were to be ignored or, better still, displaced.

With hindsight, the policy divide which came about in response to the food scarcities of the 1960s had important consequences. The large-scale hydraulic enterprises and the emphasis on mass irrigation set India on the path of unsustainable water resources management. And the other emphasis—on 'safe' drinking water supplies—overlooked the fact that rural people's need for water is dominated by livelihood and quality of life concerns, not by the threat of infectious disease. It also ignored the fact that household and environmental sanitation, rather than water supply, is the key to public health.

Village India operates according to environmental health and socioeconomic dynamics different from those of industrialized Europe and Northern America. As in other parts of the pre-industrial rural world, its disease load is not susceptible to an approach based on Western public health engineering precepts. These are only applicable in towns where virtually every household can have a septic tank or a water and sewerage connection to an efficient, centrally-managed system. That is never likely to be possible for remote communities in India, where only the grandest households with their own piped supplies and drainage systems can install modern bathrooms and flush toilets. For the typical rural family, sanitation has to be effectively 'dry'—using the barest minimum of water for pour-flushing. Instead of water closets connected to sewers, individual toilets which confine human wastes to pits or closed chambers until they are sanitized and safe are the only practicable option.

Where water supplies, drainage, and the provision of suitable means of excreta disposal, are not physically connected to a sewered system, the services and installations have to be dealt with separately from one another. Addressing them as one package under a water supply banner will never do much for public health because storm drainage and human waste disposal are thereby ignored. Water alone may even worsen the public health environment if soak-pits or other drainage is not provided to prevent standing water collecting at the standpipe or pump. Moreover, 'safe' drinking water cannot automatically confer on people healthier lives, even if no decent life can be led without a reliable water supply. Without water it is certainly difficult to wash, to keep bodies, plates, food, pots, clothes or anything else clean, or to maintain any reasonable standard of personal hygiene. But this proposition—that water is essential for a decent, healthy, and dignified life—is very different from the supposition that a 'safe' water supply will conquer infectious diseases. Many erroneous policy decisions and wrong analyses of programme outcomes have stemmed from the false assumption that 'safe' water is the key to public health.

The international community has been guilty of perpetuating this idea. The World Health Organization (WHO) asserts that 80 per cent of diseases are 'water-related', but it would be much more accurate to describe them as 'sanitation-related'. Bracketing the two areas together is helpful in one regard: unless sanitation, with its implications for environmental pollution and disease control, is linked to water supply, the risk is that no one will do anything about it. But the linkage has also obscured the need to address water and sanitation needs in rural environments with independent approaches. The 'watsan' (water and sanitation) public health agenda was given a strong push by the International Drinking Water Supply and Sanitation Decade (IDWSSD) from 1981 to 1990. India responded to the Decade by devoting considerable resources for rural water supply in successive Five-Year Plans. From Rs 5 billion in the period 1969–74, the amount devoted to water supply and sanitation rose significantly over the decade, to Rs 40 billion between 1980–5, and Rs 65 billion in 1985–90 (UNICEF 1998). Although these allocations did refer to water *and sanitation*, in reality the overwhelming share went to water supply.

Sanitation has always been a difficult subject in India. For centuries the removal of dirt from households and streets was assigned to the 'sweeper' caste. With the erosion of a system which Gandhi challenged as a structured form of human degradation, a hiatus developed. Despite all Gandhi's efforts to encourage Indians to adopt the use of sanitary pit-toilets for excreta disposal, official distaste and popular resistance to a taboo subject kept sanitation firmly in the shadows. When V. S. Naipaul wrote about the Indian practice of 'open defecation' in *An Area of Darkness*, published in 1964, the book was banned in India because it was offensive. He explained that the Indian peasant suffers claustrophobia if 'he has to use an enclosed latrine', and that the society as a whole enjoys 'a collective blindness about the practice, arising out of the Indian fear of pollution and the belief that Indians are the cleanest people in the world'. Naipaul's observation about establishment squeamishness on the subject of excreta and its disposal is borne out by the fact that there was official tardiness in addressing the growing insanitary conditions being created in an ever more crowded landscape. The rural drinking water programme, by contrast, enjoyed huge official support and political popularity. Politicians campaigned on water supply in rural areas, promising constituents that free services would materialize if their votes went their way on election day. But sanitation? No one mentioned that.

As the 1980s progressed, it began to become obvious that drinking water supplies—even if they were safe, even if they were functioning, and even if they were used—could not be instrumental in transforming public health. Specific attention was needed to be given to sanitation—both in the context of excreta disposal and for drainage of standing water—clearing of animal wastes from village paths, and regular rubbish disposal. The commitment to the Water Decade goal of 'water *and sanitation* for all' finally goaded Indian officialdom into producing some kind of systematic response to the hazardous practice of abandoning human dirt to the natural cleaning powers of the sun, sea, and soil. In 1986, with support from a variety of international donors, a national rural sanitation programme was launched. But the attention and resources allocated to it have always been overshadowed by the far more popular demand for water. Although it is true that the political games around water and the top-down delivery by the government of free water services have been a mixed blessing, at least water programmes have a lot going for them. The subject of sanitation has become more openly talked about than it used to be, and there has been some striking progress in parts of the country, but the problem of sanitary waste disposal is still a difficult nut to crack.

During the 1990s, resources continued to be poured into village water supplies: Rs 167 billion in 1992–7, and a still greater amount—Rs 420 billion—projected under the Tenth Plan period (2002–7). But an anomaly began to emerge. No matter how energetically the goal of 'drinking water for all' was pursued via expanded coverage of handpumps and boreholes, there was the appearance of running fast only to stand still. The drop in water tables all over the country caused by the uncontrolled extraction of groundwater for irrigation meant that many of the new installations functioned only sporadically or dried up in the summer months. The failure to integrate water policies meant that competition between resource usages and users was growing. In theory, drinking water supply took ascendancy over all other uses. This was stated in the 1987 national water policy and restated in the revised version of 2002: 'Drinking water needs of human beings and animals should be the first charge on any available water' (GOI 2002). In practice, drinking water has always been the poor relation because all the enhanced productivity, and thus almost all the economic and commercial possibility from the development of water resources, lies with irrigation, not with domestic supplies. The fate of the drinking water programme was captive from the start.

The development in the 1980s of a national water policy has so far done little to correct the imbalances which stem from fragmented management of the resource, from lack of rigorous groundwater regulation, and from failure to promote drought-proofing energetically in arid and semi-arid areas with high rainfall variability—even though these aims are acknowledged in the policy (GOI 2002). Meanwhile, piped water supplies in towns and better-off rural communities suffer increasingly frequent cut-offs as consumer pressures grow. Political disagreements and popular protests over water—in the Kaveri and Indus basins, in Kachchh and Saurashtra in Gujarat, in the Narmada valley, at Tehri on the Ganga, on the Shivnath river in Chhattisgarh, and in Plachimada in Kerala where commercial exploitation of water is ruining local agriculture—have become only too familiar. The people's movements formed to resist large dams and canal projects, and to protest the privatization of services and even of water courses themselves, demonstrate the fears of further livelihood loss felt by thousands of anguished marginal families in an ever more water-scarce land.

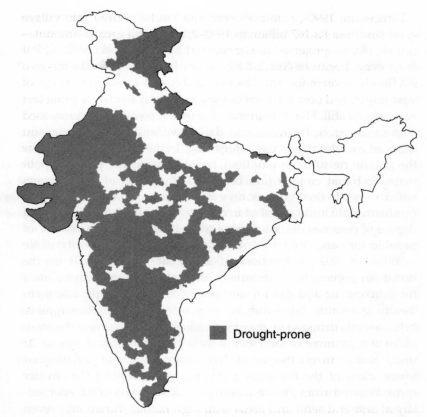

Map 1.1: Drought-prone Areas of India
Source: Central Groundwater Authority of India.

That leads us back to the subject of drought. Less rain or erratic rain, resulting partly due to climate change, has been India's trend during the past decade (Sharma 1999). According to rainfall data from Agra, Dehra Dun, and Delhi, the number of rainy days during the monsoon season in the middle and upper Ganga basin has been declining over the past 30 years, while the annual maximum temperature is on the rise (Kothyari et al. 1997). In several parts of the country—Gujarat, Madhya Pradesh, Rajasthan, and Orissa, to mention just a few—the rain gods have become increasingly temperamental. The monsoon can no longer be relied on to appear on schedule;

the rains are late, or they shower and then go away for a while, leaving farmers nonplussed about planting, or having to watch half-grown plants wither in unaccustomed heat. Elsewhere—in Orissa, West Bengal, and the North-East—sporadic rain is too often followed by floods, sweeping away barely established crops. Given India's vast population and the increasingly voracious demands of farmers and urban consumers, the trend of aberrant weather patterns, local-ized catastrophic flood, rainfall deficiency, and seasonal migration induced by water deficit, is frightening. In the end, people living off the natural resource base of land, forests, rivers, and petty entrepre-neurship based on their products—as the majority of India's people still do—need water because they have to live, grow food, engage in productive life, fulfil social and religious obligations, and enjoy some degree of personal dignity. Without this basic minimum, no one can provide for their children or sustain any reasonable quality of life.

Whatever the predilections of policy-makers and their interna-tional supporters, it is inescapable that, at community level, 'water for drinking' and 'water for livelihoods' transect. People know that 'health is wealth'. They can be very receptive to admonitions for behavioural change, once they understand the reasons and if they are given the chance. No one wants their child to become sick or die unnecessarily from a disease related to poor hygiene or pollution. But water scarcity, and the accompanying degradation of the environ-mental resource base, is a problem in another league altogether. This larger 'water crisis', in which safe and sufficient drinking water represents only one, albeit a critical, element, is becoming a real threat to many people's whole way of life.

Until that reality is factored into the entire gamut of development policies, no amount of vigour behind drinking water programmes will solve the water problems faced by India's millions of impover-ished rural families.

Unicef is a rather special international organization, which has its own perspective on social issues and on the role of water and sani-tation in human well-being. From the start of its involvement in water and sanitation in India, Unicef's motivation was to promote the health and well-being of children and women, especially those in poor rural communities. But the way in which this has been done reflects the many evolutions in its own interpretation of its mission, as well as the influence of contemporary fashions in development thinking, and the changing realities of life in India's thirsty villages.

In the very early days—the late 1940s and 1950s—Unicef's inspiration was purely humanitarian. Its principal mission was to promote the social welfare of children, often in the wake of war or disaster. Within the international community, no role was perceived for an organization concerned with children in the more muscular process of development. That task was the provenance of specialized agencies offering technical expertise such as the WHO and the Food and Agriculture Organization (FAO), and international investment bodies such as the World Bank and the United Nations Development Programme (UNDP). However, Unicef's concern with social welfare led irrevocably to social development, based on an expansion of public health and education services. As ideas of what constituted true 'development' changed, and moved closer to the concept of improvement in human well-being and less exclusively concerned with macroeconomic performance, Unicef's sense of its role also changed. It had the advantage of being close to the ground and intimately involved with the government-led programmes its assistance supported; at that level it was clear that genuine development required social, as well as economic, investments. By the 1960s and 1970s, Unicef's programme straddled two different mindsets: concern with children and families as its raison d'être, but by means and methods which matched priorities in national development plans. It began to shed its purely humanitarian image, cast off its international subservience to the technical expertise of WHO and FAO, and claim its own development role.

In India, water supply was the route by which Unicef stepped over from small-scale humanitarianism to major developmental input. This came about in response to a variety of circumstances. During the 1950s and 1960s, Unicef had provided a trickle of support to water projects for small-scale community schemes, and facilities in health centres and schools. The departure into a much more significant water supply involvement grew out of its response to the 1966–7 Bihar famine emergency. As part of the national and international relief effort, thousands of shallow wells were dug to provide drinking water and irrigate land. For hard rock areas, Unicef imported high-speed drilling rigs so that inaccessible supplies of water deep underground could be reached as fast as possible. This story, as well as its sequel, is the subject of Chapter 2 of this book.

The decision in 1969 to support the new national rural drinking water programme by supplying 125 state-of-the-art drilling rigs marked a turning point in the story of Unicef's cooperation in the sub-continent. At the time, this was the largest grant, and technologically the most ambitious, that Unicef had made anywhere in the world. Within Unicef itself, it was highly controversial. It was a daring venture since it involved introducing, on a significant scale, a novel and sophisticated technology into the Indian environment. In a very real sense that decision was the platform from which everything Unicef has subsequently done in India was launched, not just in water supply, but in the whole range of its cooperation for health, education, nutrition, women's welfare, and the rest. That grant caused Unicef to be seen as a serious development partner at the highest levels of central and state government. No longer a mere provider of child welfare and nutrition supplements, Unicef had identified itself with a problem at the heart of all development in India: the provision of water, the essential input for life and health, in remote and neglected corners of the land.

Today, there would be no question of Unicef making an equivalent investment in capital equipment for any programme, let alone one that was experimental. But this was the heyday of belief that the key to economic and social development lay in relatively grand inputs of technological fix. Unicef was no more immune to this idea than anyone else. In fact, Unicef believed that it was specially endowed with the capacity to bring the marvels of modern science and technology to its child-focused mission: the investment in Indian water supplies fitted exactly into its desire to be regarded as a development, as opposed to a purely humanitarian, organization.

Of course, the assistance Unicef offered was not on the macro-scale of multimillion investments in, for example, large dams and other infrastructures supported by such external agencies as the World Bank. And its intention was geared to a very much more modest level of input than rearrangement of a vast river basin or a trunk pipeline to supply a metropolitan centre: one-by-one installation of 12,000 boreholes a year in disadvantaged rural communities. But the proposal was decidedly out of the ordinary in Unicef terms. One of the reasons it landed on Unicef's desk was that there was no other international organization interested in rural water supply that the Indian government could turn to—at least, not if serious amounts of equipment were required, and not just the technical advice UN specialized agencies could offer. Unicef, with its exceptional UN mandate to provide material assistance for 'the benefit of children' and its concomitant experience in supplies procurement and delivery, was the only possibility. So the proposal, and its acceptance, led to the incongruous situation where Unicef, which was not a technical organization, became both the medium through which high-speed hard rock drilling was systematically introduced into India, as well as the pre-eminent external support agency for the programme. This, in turn, affected Unicef's worldwide programme of assistance: its own expertise in water supply grew to respond to the needs of this programme, and was then put to use elsewhere in the world. During the late 1970s and early 1980s, water supply was for a time the area of Unicef's worldwide programme absorbing the largest proportion of funds.

Never an organization to rely on perfect planning to prevent disasters from occurring, Unicef adopted not the 'far consultants visiting sporadically from abroad' pattern of involvement, but the 'hands on, let's solve this locally' approach which has, uniquely within the UN system, always been its way. Unquestionably, the full implications of the commitment in India were not initially apparent to Unicef. But once up to the waist in any major programme, there is an inevitability about plunging further in. In its early years, as is invariably the case in a major development enterprise, the programme threw up a deeper set of needs and a wider range of problems than had been visible when it began. Not to face up to them would have seemed irresponsible, not to say callous. When the US$ 600 million Minimum Needs Programme (MNP) was under discussion for inclusion in the 1974–8 Fifth Five-Year Plan, Unicef committed itself to support an expansion of rural water supply by a further investment in capital equipment. These large—for Unicef—commitments had the effect of stamping a special character on the programme. The organization's investment had to be protected, which led to a build-up of its own in-house technical expertise. Engineering professionals were hired by Unicef to liase with rig manufacturers and provide technical back-up to the states. As each technological innovation to the programme threw up the next generation of hitches, Unicef's team was there to help resolve them.

For nearly 20 years, the technological agenda defined the nature of Unicef's involvement. The Unicef team led the first systematic effort to develop a new handpump for community use in hard rock areas (the subject of Chapter 3). In time, the degree of concentration on engineering excellence and technical back-up began to sit uneasily with other directions in which Unicef, as a child-focused social development organization, was going. But with a central and state government wind behind it, and with all its own in-built momentum, it was hard to wrest the programme away from the comfortable embrace of the classic supply- and technology-driven model. Even when Unicef joined the group of donors encouraging the CPHEEO to invest in a major sanitation programme during the international Water Decade (see Chapter 4), its emphasis was as technologically focused as the rest.

Within its own parameters, the rural water supply programme was a huge success. But it has never been without its critics, even from the early years. For members of the NGO and activist community, it was deficient in transferring to the agency of the state the responsibility for water supply provision, and eclipsing tried and effective solutions—dug wells and water harvesting structures in particular—which were within the capacity of poor communities to install and manage. According to this view, the economics of technology-led solutions, such as high-speed drilling, placed this kind of response to water scarcity beyond the internal means of low-income communities. It also sapped local ingenuity and self-belief, created dependency on inefficient and under-funded government structures, and was ultimately unsustainable. In Unicef's view, however, the technology was a relatively low-cost response using modern scientific techniques to provide basic services at the community level—a bridge between unsafe traditional solutions such as dug wells, and the macro-schemes for metropolitan areas which are unaffordable for rural populations.

Although this idea has much validity, it does not respond to the very real problems of community usage, management, and maintenance of the water supply installations provided *for* rather than *with* villagers, without consultation about how best their problems could be solved. The solution was not always useful or acceptable, and sometimes it was rejected by simple neglect. The high rate of breakdown, and lack of effective systems of repair began to be addressed in the late 1970s; but 25 years and many policy evolutions later, issues of sustainability and affordability at community level are still far from resolved (see Chapter 6).

In recent years, the received international wisdom has changed: community water supply schemes should be 'demand-responsive' with some of the costs of installation and all of the costs of maintenance passed on to the users. However, it is far from easy in India to abandon the 'free service' ethic of the past, nor is it politically popular. Meanwhile, the pressures on groundwater have grown to the point where, in many hydrogeological settings, there is no realistic alternative to mechanized drilling for reaching deep into aquifers. Yet the handpump–borehole technology is still beyond the means of many low-income communities, even to pay for major repairs, let alone for installation. Their lack of consumer power and resources complicates efforts to decentralize service management. Unless alternative options can be found, there is today a risk that better-off communities will get mechanized boreholes, and the poorer farmers, the *adivasis* (tribal people), and the landless, nothing at all.

During the 1980s, another 'software' aspect of the programme also began to preoccupy Unicef. This was a time at which the organization became tightly focused on issues of child survival, through the promotion of disease control in the under-fives. The disappointing results of water supply schemes in the health context, not only in India but elsewhere, led to a downturn in organizational interest in water supply generally. Unicef's water teams were forced to consider how to ensure that the programmes they were involved with led positively to the health gains which had earlier been so confidently predicted. Not only did this lead to heightened organization interest in sanitation, but also to establishing specific links with disease control and to the promotion of hygienic behaviour. Unicef began to back health and hygiene education in schools and *anganwadis* (community childcare and women's welfare centres at the village level), through NGOs, and via government programmes in states such as Orissa, Andhra Pradesh, and Karnataka whose authorities were positive to this approach. It also played a leading role in the eradication of guinea worm from India, achieved in 2000 (see Chapter 5). And it focused on cleanliness—especially the washing of hands—to break the transmission route of diarrhoeal infection. However, gains in this context still remain elusive. Even till as late as in 2003, around 400,000 Indian children are losing their lives every year to some kind of bacteriological or viral disease that attention to hygiene might have prevented.

Thus, at different times over the 35-year old history of the Unicef assistance programme, organizational faith has been placed in different aspects of the water and sanitation package to bring about the necessary breakthroughs. From an initial, almost exclusive, emphasis on technology and hardware, there has been a sea change towards today's almost exclusive emphasis on software interventions: community consultation and management of systems, and the need to inculcate hygienic attitudes and behaviours. Ironically, Unicef's most prominent focus today is in the schools where it first began promoting water and sanitation more than 40 years ago. Today's aspiration is that if school children are taught to practise healthy living, if no girl or boy wants to marry into a family without a toilet, then surely an environmental sanitation and family health transformation will occur.

Hanging over the potential for better health and more dignified lives, however, is the rising threat of environmental pressures and water scarcity. Unicef rightly takes pride in the vital role it played in the transfer of deep-borehole drilling technology to India in the name of domestic water supplies for children in poor communities; but it was slow to appreciate the long-term implications for the water table unless measures for recharging aquifers were simultaneously put in place. Here is an example of a basic development truth: technical solutions to development problems provided with the best of contemporary evidence that they will be suitable for their settings, in course of time themselves produce technically suspect outcomes that were never anticipated. While hindsight is a wonderful thing, and no purpose is served by rounding on those who did not read a technological crystal ball, it is not to their credit if those who played a part in a much-applauded drama of development success walk off the set, ignoring what has since come about partly as a result of their intervention. In the context of water supply programmes, there has been a tendency for Unicef to draw the wrong conclusion: stay away from technical interventions and stick to software alone. But both are vitally important.

Those who played a part in the introduction of borehole technology to India need to look at ways to preserve their own inheritance before it goes the way of the hydraulic efficacy of India's long-distant past. In this context, it is today the recharge technicians, catchment experts, community management practitioners, and quality testers who are leading the pack. The gatekeepers of national and international water wisdom need to help bring community-based initiatives for rainwater harvesting and aquifer recharge into the policy mainstream. While Unicef is not inactive in these areas, the role that it could have played in drought mitigation has not been fulfilled to the extent that it might have been (see Chapter 7).

Focused narrowly though any programme may choose to be, in the case of water and sanitation, there is no way to avoid the wider context—the economic, environmental social, and political dimensions—which both drives programmes forward and lands them in backwaters and on shoals. In the climatic uncertainties of the twenty-first century, in the growing competition in India over scarce freshwater supplies, in the increasing use of market mechanisms to decide allocations, in the growing presence of chemical and bacteriological pollution, there will be a need to protect the rights of the weakest and most vulnerable members of society—the children. Unicef's long experience in India, and the evolution of its water and sanitation programme to meet problems as they emerged, contains many useful lessons. Whatever the needs in other areas—health, education, women's status, income generation, household food security—water is essential to life and health and fundamental to every productive human activity. On the lessons of the past, the future must be based.

REFERENCES

Agarwal, Anil and Sunita Narain (eds) (1997), *Dying Wisdom: Rise, Fall and Potential of India's Traditional Water Harvesting Systems*, 'State of India's Environment' chapter 4, CSE, New Delhi.

Black, Maggie (1987), *The Children and the Nations*, UNICEF/MacMillan, New York/Sydney.

Child's Environment Programme, UNICEF/Government of India, Master Plan of Operations, 2003–7.

Glieck, P. H. (1998), *The World's Water: The Biennial Report on Freshwater Resources*, Island Press, Washington, D.C..

Government of India (2002), *National Water Policy*, Ministry of Water Resources, paras 7, 8, and 19, April.

————— (2001) Census, Tenth Plan Document.

Jaitley, Ashok and Raj Kumar Daw (1995), *Contribution of Voluntary Organisations in Rural Drinking Water Supply and Sanitation Programmes in India*, paper sponsored by UNICEF, Delhi and presented to the Water Supply and Sanitation Collaborative Council in Barbados.

Jamwal, Nidhi and D. B. Manisha (2003), 'The Dark Zone', in *Down to Earth* magazine, CSE, 8 April.

Khilnani, Sunil (1997), *The Idea of India*, Penguin, quotes taken from speeches of Jawaharlal Nehru.

Kothyari, U. U., V. P. Singh, and V. Aravamuthan (1997), *An Investigation of Changes in Rainfall and Temperature Regimes of the Ganges Basin in India*, Water Resources Management, Vol. 11, No. 1, pp. 17–34, February.

Meller, Helen (1990), *Patrick Geddes: Social Evolutionist and City Planner*, Routledge, London.

Nigam, Ashok, Biksham Gujja, Jayanta Bandyopadhyay, and Rupert Talbot (1998), *Freshwater for India's Children and Nature*, UNICEF and WWF, New Delhi.

Sharma, Devinder (1999), 'The fight for food', in *New Internationalist* magazine, December.

UNICEF (1998), *A Time Line*, UNICEF Water and Environmental Sanitation Section.

Verghese, B. G. (1967), *Beyond the Famine*, Bihar Relief Committee, New Delhi.

Viksat (2003), Personal communication, Ahmedabad.

Ward, Colin (1997), *Reflected in Water: A Crisis of Social Responsibility*, Casell, London.

2
Finding Water for
Water-short Communities

If unpredictable rainfall is an annual bane for the Indian farming family, the rock structure under the soil on which they farm has been an even more intractable obstacle.

India's geological make-up is complex, but can be broadly classified into three zones: the soft, alluvial tracts of the Indo-Gangetic plain; the coarse, bouldery sediments of the Himalayas; and the hard rock formations—basalts and granites—that are found in 80 per cent of the country and predominately in the Deccan plateau (Dey 1968). These hard rock areas are India's geologic curse. They are of great antiquity and the underlying cause of the country's proneness to drought. In younger, more broken-up sediments and alluvial areas, groundwater aquifers are like saturated sponges that leak their contents into springs and wells. These aquifers are both porous and permeable, and water can pass through their silts, gravels, and shales relatively unimpeded, lodging in their pores. They are easily recharged and can often sustain heavy pumping without much depletion.

Hard rock aquifers behave quite differently. Rainwater filters with difficulty down to the bedrock below, filling whatever fissures, joints, and cracks it can find. It takes decades for shallow fractures to be filled; centuries or millennia in the case of deeper ones. As a result, aquifers in India's hard rock areas are meagre and miserly; they have little water to give and what they have, they relinquish reluctantly. They are also elusive. Even with the aid of modern geophysical

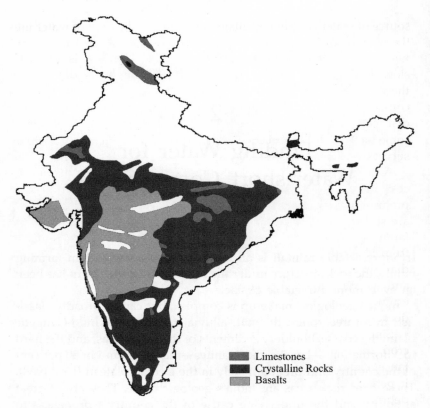

Map 2.1: Hard Rock Areas
Source: Central Ground Water Board, February 2004.

equipment their detection makes waterwell prospecting in hard rock areas as much an art as a science.

Until recent decades, that kind of prospecting for water was unnecessary. Enough water could be found within the mantle of soil and weathered material lying above the bedrock, the stratum known as the 'overburden'. Wells were traditionally dug within the overburden. In drought-prone areas of peninsular India, the overburden is invariably shallow. When hard rock was reached, well-digging ended or else the rock was blasted out using gunpowder. Holes to take the blasting charges were laboriously hand-bored with hammer and chisel. In the case of these dug wells, many of which constitute a key source of water supply for villages even today, the flow of water into the well depends on whether the well is close to rivers or tanks, and how much water is contained in the mantle. The flow is generally slow, so the size of the well is important. The larger the diameter of the well, the greater the inflow. Such wells take months or years to construct. In parts of southern India, such as in the Coimbatore district of Tamil Nadu, where hard, crystalline granites predominate, some wells have been blasted and deepened to 50 metres or more by successive generations of the same family.

Although the old technological approaches for groundwater access have long been superseded within government-sponsored programmes by borehole drilling, these practices for well-deepening are still pursued in drought-prone areas. This is often done at the family's or community's private initiative, sometimes with support from NGOs but not necessarily. It is worth re-emphasizing that Indian villagers have, over the centuries, exhibited considerable skill and wisdom concerning water. The idea of depending on the government to provide water, however ingrained it has become in recent decades, is in fact relatively new.

Since an Indian farmer's income is dependent on the success of his crop, the health of his well has always been of paramount importance. He invests in his well, protecting it with a parapet, deepening it periodically to make sure it doesn't run dry, and cleaning it of debris every year to prevent obstruction to water flowing in. But whatever his determination to preserve his greatest asset, he cannot guard against over-withdrawal of groundwater by a neighbour with ambitions identical to his own: to preserve his well, irrigate his crop, sell his produce, and feed his family. Heavy extraction from one well affects water levels in wells in the same vicinity. Since colonial times, groundwater in India has been legally recognized and exploited as a private, not a public, resource. Whoever owns the land on which the well is sited is legally able to draw whatever volume of water he likes, irrespective of the fact that the groundwater aquifer is not contained by any such boundary.[1] So the farmer with the deepest well

[1] This legal position may be beginning to change. In late 2003, the Kerala high court broke with legal precedent and ruled that groundwater was a resource belonging to the entire society, and that no more than a 'reasonable' amount could be extracted by a landowner.

rules. Only recently have some states begun to try and regulate groundwater extraction by new legislation.

In the 1960s and 1970s, as pressure on groundwater began to grow in hard rock areas, richer farmers began installing power pumps on their wells. The less well-off farmer, dependent on his bullocks and the leather buckets they pulled slowly from his well to tip into his irrigation furrows, was helpless when his well ran dry. More ominous still were the deep, drilled borewells that some farmers began to install. These threatened the smaller farmer's lifetime investment. And year by year, the water level in his well dropped alarmingly. In some cases, it vanished altogether.

Until the late 1960s, the only machines used for drilling for water in hard rock areas were cable-tool percussion rigs. These simple and effective machines are still the rig of choice in India's boulder sediments and softer formations. A hardened steel drill bit attached to a cable is repeatedly raised and dropped, pulverizing the rock beneath. The chips and slurry into which it is rendered are removed from the hole by a bailer. Mounted on truck or trailer, these 'bash and splash' rigs are inexpensive to buy, run, and maintain. But in hard rock, they are slow. In very hard rock, they typically drill only a metre a day. In places where depths of 80 metres and more are required and a sense of timelessness is attached to the process, they sometimes only manage one or two wells a year. In the India of the 1960s, there was nothing to challenge this pace of drilling.

The person who originally changed this was a Church of Scotland missionary, an agricultural engineer based at Jalna in Maharashtra. John McLeod was on leave in Yorkshire, England, in 1963 when he happened to pass a limestone quarry and stopped to inspect a drilling machine throwing up clouds of dust. This mobile rig was using compressed air to drive a cylindrical steel hammer with a tungsten carbide tipped drill bit into the hard limestone, and simultaneously to flush out the rock cuttings, at a pace McLeod could not believe. This was his introduction to the Halco Tiger, a drilling machine that used a down-the-hole (DTH) hammer and whose technical advantages over cable-tool rigs in terms of hard rock penetration and speed belonged—as far as Macleod and others like him were concerned—to the category of 'wonders of the modern world'. He thereupon determined to use these rigs to speed up waterwell production in the villages located in a hard rock area of Maharashtra where he was operating. Having garnered advice, support, and funding from the British charities Oxfam and War on Want, McLeod brought the first Halco Tiger into Maharashtra in 1965. The original purpose was to help small and impoverished farmers to increase their agricultural yields. It is worth noting that the high-speed rig was not initially perceived as a solution to drinking water problems.

In both 1965 and 1966, there was a serious failure of the monsoon. Drought descended over several states, of which the worst affected was Bihar. In the northern part of the state, occupying the alluvial flood plain of the Ganga, the wells feeding the dry-season crops could easily be deepened. But in the south, where crystalline granite underlies a thin mantle of overburden, crisis prevailed. Across the Chota Nagpur plateau, the water table dropped below the floor of the shallow wells. Deepening them was seriously problematic. The rock presented an impenetrable barrier to hand tools, and blasting was slow, crude, and dangerous. Water began to disappear, crops and livestock perished. In districts such as Ranchi, Singhbhum, Palamau, and Hazaribargh, the autumn crop of 1966 produced only a quarter of its usual harvest while in some places it produced next to nothing. Because this was the second year of drought, there was no grain, either to eat or sow. It had all been consumed. By late 1966, the prospect of famine deaths and mass out-migration to relief camps began to loom. Around 30 million people faced starvation, and with memories still vivid of the 3 million who had died during the West Bengal famine of 1943, an outcry began. A Bihar Relief Committee was set up under the Gandhian politician, J. P. Narayan. Its twin key concerns were the procurement and distribution of food rations and the expansion of water sources. In November, Prime Minister Indira Gandhi declared a national famine emergency, and appealed to the international community for help.

The Halco Tiger, whose speed, din, clouds of dust, and spectacular ability to punch through hard rock had already won popular appeal in Maharashtra, was the machine of the moment. If the bore was successful, a new water well could be provided within two days. The missions in Jalna, Vadala, Indore, and other places in Maharashtra sent across some of their drilling rigs to aid the Bihar relief programme. When the GOI approached Unicef for assistance, it took advice from the NGOs with Halco Tiger experience, and air-freighted nine of these rigs from the UK, putting them to work in southern Bihar. The enterprise was in the best tradition of 'all hands to the rescue'; and this was an era when heroic missionary effort from the ex-colonial alma mater was still part of standard emergency response practice.

Not all of the boreholes were successful. Drill bits wore out much faster in the abrasive granite of Bihar than they did in the basalt of Maharashtra or the limestone of northern England. Equipment failed in temperatures which soared over 40°C and stayed there for weeks. The Halco Tiger was insufficiently rugged to cope well in these conditions, but impressively, Bihari mechanics kept them going. Within two months, they had pounded through soil and rock to bring water to 250 villages whose inhabitants had been faced with imminent evacuation to drought refugee camps (Black 1990). Whatever the shortcomings of this crash drilling programme, the miracle of water from hard rock in a matter of days deeply impressed the government and the international donor community, especially Unicef.

Out of this emergency relief action grew a national programme which marked a turning point for government involvement in water provision for village India. Beyond its significance for the sub-continent, the programme was also the first large-scale engagement by an international or UN organization in the operational side of drinking-water supplies in rural, and quintessentially pre-industrial, settings anywhere in the world. International water experts regarded the India programme as a test-tube for similar ventures, then under consideration in other countries. That a number of technical challenges faced this new national and international cadre of public-health engineers was well understood, but just how many and how complex they would be remained a matter of conjecture.

The experiences with DTH hammer rigs during the Bihar drought helped promote an air of enthusiastic zeal around the prospects for providing water-short villages with what they most needed—dependable wells. In the late 1960s and early 1970s, in a spirit of technological optimism, a host of NGO water development projects sprouted across the hard rock areas of the country: at Indore and

Betul in Madhya Pradesh; Coimbatore and Madurai in Tamil Nadu; Jalna, Vadala, and Sholapur in Maharashtra; Bangalore and Hubli in Karnataka (then Mysore); Ajmer in Rajasthan; Hyderabad in Andhra Pradesh. Drilling equipment was supplied by Oxfam, Christian Aid, War on Want, and other charities under the auspices of Action for Food Production (AFPRO), an NGO umbrella organization set up in Delhi in 1966 to provide technical advice and spare parts. So powerful was the image of the 'Tiger' coming to the rescue of India's villagers that the most famous pictorial weekly in the world, *Life* magazine, ran a photo-feature of the drilling project run by a larger-than-life NGO personality, Vincent Ferer, in Manmad, Maharashtra (Jaitley and Daw 1995).

Ferer was just one of many remarkable characters involved with the water projects of the time, including John McLeod and his geophysicist colleague, Chris Wigglesworth, at Jalna in Maharashtra, Charles Heineman at Madurai, and Arden Godshall at Bangalore. All were either missionaries or employed by missions, and were dubbed 'mechanics with a message' by the motley collection of young British, American, Dutch, and Danish volunteers sent out to work with them. These water projects were not all exclusively expatriate ventures, although in most cases expatriates were their leaders. There were enthusiastic young Indian hydrogeologists and engineers too: Vishwas Joshi at Betul, Raj Kumar Daw at Vadala, Bunker Roy at Ajmer. Energy, zeal, a practical bent, disregard for the discomforts and hazards of the drilling life, and a healthy disrespect for the book of rules were the hallmarks of these young pioneers' attempts to get things done. Many of these water mechanics of the 1960s and 1970s, both Indian and expatriate, went on to make water provision their life-long profession. With their exposure to people, local cultures, and awkward ground realities, they brought down-to-earth common sense to air-conditioned meeting-rooms where theoretical solutions to problems facing poor people too often held sway. Some were to depart radically from the unquestioned belief in the rigs and boreholes that originally inspired them. But those misgivings came later (Jaitley and Daw 1995).

As far as the government was concerned, the Bihar famine had been a searing experience, from which starvation deaths on a scale of millions rather than thousands had narrowly been averted. There was a determination that drought must never again be allowed to lead to famine. The performance of the high-speed drilling-rigs had been impressive, and the authorities began to look to these—and to Unicef—as the means of future salvation. The Fourth Five-Year Plan (1969–74) was in preparation, and the idea of adopting DTH hammer drilling for mainstream water supply provision instead of for belts-and-braces emergency relief had great appeal.

Since the 1950s, the CPHEEO and state water departments had been involved in the provision of community water supplies. But not by this method and not on any significant scale. DTH drilling was by now well established among water-conscious missions and NGOs, and in their own small way their efforts were already quite wide-spread. They had begun to network under the auspices of AFPRO and local NGO federations such as Action for Agricultural Renewal in Maharashtra (AFARM) in Ahmednagar (Jaitley and Daw 1995). But they did not attempt to 'sell' their approaches to the government or to local institutions, whose operations they typically regarded as completely distinct and unconnected to their own. Thus, within the context of regular government-sponsored rural development activity there had been no systematic deployment of the new water well drilling technology up to this point.

The essential groundwork for a major drinking water supply programme had been laid in 1962. The Ministry of Health had un-dertaken the first survey to identify 'problem villages'—a definition then used exclusively to describe communities that were water-scarce, later broadened to include other types of water problem (Black 1990). This exercise established the basic strategy for India's rural water supply programme, a strategy whose underlying principles have remained intact ever since. Criteria were established for identifying 'problem villages'. These were villages without a year-round water supply within 1.6 km distance in the plains, or in hilly areas, at 100 metres of elevation from their supply. Inventories of 'problem villages' were then developed, block by block, district by district, and state by state. Altogether, 153,000 such villages were identified at this time. The installation of water systems in these villages had then become the target of the national programme. However, the majority were in hard rock areas, and before the advent of high-speed drilling, the authorities had no effective way of helping them.

By 1969, the Bihar drought and growing concerns about dropping water tables led to a plan to develop water sources in hard rock terrains—classified as 'technically difficult areas'. Unicef was approached for assistance in procuring the fastest, most up-to-date and appropriate hard rock drilling equipment. For Unicef, this was an unusual—not to say unprecedented—kind of request. Its staff were very aware of the problems encountered during the drought relief programme in Bihar: expensive hammers lost in boreholes, a shortage of trained operators, a lack of repair facilities and spare parts. The Halco Tigers had also suffered a high attrition and breakdown rate: their tigerish qualities had not been conspicuous on Indian roads or in Indian climatic or geologic conditions. There were many in Unicef who believed that sophisticated drilling equipment was not part of its development agenda. There were better options and, that too, at a lower cost, such as health care and nutritional supplements for women and children. Some believed that India could not manage the technology. It would be better to supply hundreds of cable-tool percussion rigs, slow though they were, rather than the faster, more complex, rigs.

However, others took the view that the supply of drinking water was not only within Unicef's mandate, but that in a country which already had a large industrial sector, modern technology was the way to provide it. Unicef's proposed investment was US$ 5.9 million—an insignificant figure by the standards of today. But its sanction required a decision by Unicef's Executive Board in New York. Initially, it was rejected on the grounds that the highly technical nature of the equipment and the risks associated with the expenditure could undermine the credibility of the organization. However, following an assessment led by Martin Beyer, a Swedish hydrogeologist, the views of the pro-water lobby ultimately held sway. A tripartite agreement was signed between the GOI, Unicef, and WHO for the 1969–74 Plan period. This facilitated the systematic introduction of high-speed hard rock drilling into the Indian public health engineering establishment and launched Unicef into its pre-eminent role as India's main external support agency in rural water supply.

The rationale for Unicef's agreement to support the 'accelerated' rural water supply programme is very significant. The thrust was safe drinking water, to pursue the goal of improved public health, particularly of children. In spite of the needs of the Indian farmer for water to irrigate his crops during the dry season, without which his family's food supply would be threatened and children's and women's well-being jeopardized from another direction, Unicef's concern was limited to water for drinking and domestic purposes. Indeed, if there had been any mention of agriculture during the debates surrounding the proposal, it would have stopped dead in its tracks. Some advocates of applied nutrition were keen to support domestic water supplies for kitchen gardens as an adjunct of family food supplies, but nutrition programmes were then seen as an adjunct to health in the Unicef perspective (Black 1987). This did not accord with the way community water resources were traditionally viewed, either by villagers or by previous government policy.

Given their multiple needs for water, including water to irrigate their crops, villagers in India tend to view the water resources available to them holistically. What they need is water, plain and simple. In many parts of the country little differentiation is made between water for drinking and domestic purposes and water for cultivation. Certain wells and other sources may be favoured for drinking because of their taste or perceived purity, or because they can be more conveniently accessed. Until the advent of the 'problem village' with its exclusive focus on defining water scarcity in terms of water for drinking, government programmes for village water supply had not made this distinction either. But in the late 1960s, influenced on the one hand by the Green Revolution and its emphasis on large-scale irrigated agriculture, and on the other by a new 'water for health' ideology promoted by Unicef and WHO, government policies towards water were for the first time compartmentalized.

The long-term implications of this division along sectoral lines were not then perceived but they were to be profound. In fact, the idea that they were promoting a departure from the norm did not occur to Unicef, whose new water professionals were schooled in Western public health engineering traditions, where domestic water supplies have no livelihood context and are almost exclusively about washing, cooking, and drinking. No one can quarrel with the primacy of water for drinking. Water to drink is indisputably essential for human and livestock survival. But a policy which neglected other basic water needs, and failed to integrate requirements for agriculture and requirements for health has become, in more recent times, an albatross of terrifying proportions. Such a crisis had not been anticipated at the time. There was a head of political steam behind village drinking water supplies and, after initial self-doubt, Unicef stood ready to serve it.

The decision to procure 125 DTH air hammer rigs was taken by Unicef in 1969. The plan envisaged two types of machine, some capable of drilling 4½ inch diameter bores for handpump supplies, others of 6 inch diameter for power pump installation in selected areas, mostly to depths of 150 feet but some as deep as 500 feet. Some conventional heavy cable-tool rigs were also sought for drilling in the bouldery formations of the Himalayan foothills. At a drilling rate of around 100 boreholes per rig per year for the DTH hammer machines, this equipment was intended to provide a good quality water supply for 12,000 villages in 11 states annually, covering the needs of a total of around 9.2 million people (Black 1987).

From the outset of this investment, there was a determination that enthusiasm for technology should not land Unicef in embarrassment. The organization did not wish to find itself explaining why very large items of expensive equipment were standing idle, parked at the side of the road, or waiting at some workshop for spare parts or for someone to explain how to use them. This was the fate of drilling rigs and other costly items of modern agricultural equipment provided under aid budgets to some parts of the developing world, where donor organizations were over-ambitious about the pace of progress, and local administrations unable to absorb and use machinery effectively. A drilling rig is expensive. Typical costs in 1970 ranged from US$ 50,000 to $ 150,000. But its purchase is only the start of the process whereby the rig is put to drilling wells. That requires a sophisticated operational and maintenance set-up: trained crews, fuel and lubricants, a supply of spare parts, workshops, hydrogeological information, logistics and route information, and carefully planned schedules to employ the rig, in the formations that match its design criteria. Funds are needed to provide all this over a rig's working life.

Investment in a drilling rig, therefore, carries with it a lifetime obligation of management, nurture, and infrastructural support.

The chances of India being able to do this, as Unicef's local chiefs had persuaded the sceptics, were good. India had all the ingredients necessary for eventual self-sufficiency in water-well technology. The new drinking water programme was driven by political aspirations and was, therefore, financially well-endowed from within the country: the programme had first call on a sizeable allocation of the Fourth Plan outlay. In addition, important bilateral donors, principally SIDA, the largest contributor to Unicef at the time, had decided to back the programme and were anxious for it to succeed. There were, too, vested industrial interests at stake—those of international drilling equipment manufacturers. Other major assets were India's engineering expertise, a bureaucracy decentralized to the district level, a well-developed infrastructure, relatively good communications, and an extensive rail and road network. At the state level, there were ready-made implementing bodies in the shape of Public Health Engineering Departments (PHEDs) or Water Boards, with their tiers of engineers, mechanics, and drillers-in-waiting. Finally, India had a strong manufacturing base with the potential for replicating imported technology. There were undoubtedly risks involved in the programme, but there was also a lot going for it.

Once the initial decision to go ahead was taken, the next question was the selection of equipment. Three internationally respected companies were selected: Atlas Copco of Sweden, Halco (the Halifax Tool Company) from the UK, and Ingersoll Rand from the United States (US). By using several rather than one company, prices would be competitive and no monopoly of supply for a programme expected to expand substantially could be obtained by one company. All three companies had a manufacturing base in India, and could provide the necessary technical support.

Together with support trucks and spare parts, the new rigs were delivered to various states during 1970–4. Since most of the equipment was landed there, Mumbai (then Bombay) became the base for a stores and warehousing operation for spare parts and back-up. These early years were a period of experimentation with equipment in different hydrogeological settings and with newly trained crews, and of constant interaction with the various companies on technical improvements and specifications. The smaller machines, those supplied by Atlas Copco and Halco, drilled the smaller diameter 4½ inch bores for the handpump programme, and were capable of depths of 200 feet in hard rock. Halco, in the light of the Bihar experience, updated the Tiger so that it had greater cooling capacity and the luxury of road springs. Atlas Copco, a company which had been firmly of the 'no sophisticated rigs for India' school of thought, had originally been reluctant to sell Unicef a DTH hammer rig. But when told that it was a DTH hammer rig or no rig at all, they produced a new light-weight rig and hammer especially. This had some strong advantages over the Tiger, the competitor for the same conditions: it was run entirely on air, and its design criteria focused on lightness, simplicity, and reliability. For 6-inch-diameter bores, the star performer was the Ingersoll Rand truck-mounted Trucm-3, a rig that had been highly successful in the US and was capable of drilling to a depth of 600 feet. The first of these rigs to be brought into India had arrived in 1968, courtesy of AFPRO. Arden Godshall of Action for Water Development, Mysore, achieved such spectacular results and demonstrated the virtues of his machine so impressively that Unicef hired him on the spot as their first master driller.

Gradually the new programme began to get underway. At the central level, the responsibility for giving overall policy guidance to the state bodies in charge of water supply lay with the CPHEEO, which came under the authority of the Ministry of Health and Family Planning.[2] The state bodies—PHEDs and Water Boards—were responsible for the actual drilling programme and all its ancillary activities: geological survey, site location, water quality testing, and equipping the boreholes with handpumps. These state bodies were variously constituted and came under the authority of different state sectors. All needed to upgrade their technical and managerial competence to use the new rigs effectively. So the state level was identified by Unicef as the main target for building capacity and human resources in the first programme phase. Three master drillers were

[2] Until the Fifth Plan period, 1974–8, when the Ministry was separately established, the full remit of the Ministry was Health, Family Planning, Works, Housing and Urban Development. The rural drinking water supply programme was located in the Health Ministry.

recruited to work closely with the PHEDs, conduct technical and administrative courses for supervisory staff, and train drilling rig operators and mechanics. They provided feedback on equipment performance, so that the engineering professionals newly recruited by Unicef in New Delhi could discuss with the manufacturers modifications based on local conditions to be incorporated in subsequent equipment orders. They also set up a rig-monitoring system to record performance against basic parameters—number of boreholes drilled, well depths, and success rate, rig by rig.

During 1970–4, the new technology proved itself beyond doubt. The speed of borehole production more than compensated for the additional capital and recurrent costs of the sophisticated rigs, making the price per borehole cheaper than traditional methods. Some spare parts were already being manufactured locally and signs were developing of a fledgling drilling industry. Although it was recognized that 125 rigs could not improve the water problems of more than a fraction of the country's problem villages, the hope was that the equipment and technical training would give each state a nucleus on which to base future programme activities.

As preparations got underway for the next Five-Year Plan (1974–9), a Minimum Needs Programme (MNP) with an investment of US$ 4285 million was proposed, in which drinking water supplies were to be a major component. A new government survey of rural water supply problem villages had been completed (1972) and the criteria had been more completely defined. Water scarcity was still the first consideration: any village more than one mile distant from or more than 50 feet above its water source was classified as 'problem'; villages where cholera and guinea worm disease were endemic, and those whose wells were saline or chemically contaminated with fluoride or iron, were also included. These new criteria meant that the figure of 'problem villages' came out very close to the original of ten years before: 150,000 (Saxena, quoted in Agarwal 2000). Unicef planned to spend US$ 9.3 million on water supplies under the MNP—the lion's share, as before, on drilling equipment. However, Unicef's input was to be more selective, emphasizing special purpose rigs. From this point on, while the main programme thrust was 'problem villages', Unicef focused on 'problem areas' within the programme. Many of these surfaced as the programme advanced; others, such as hard-to-reach villages and exceptionally difficult geological conditions, had always been there but were awaiting technological solutions.

A 1974 evaluation of the Unicef-assisted programme was the prelude to the new investment. This was conducted as a matter of routine. But to the horror of the responsible officials, spot surveys in Tamil Nadu and Maharashtra revealed that 75 per cent of the handpumps on the new boreholes were out of action. After all its apparent success, the high-profile, high-tech drilling programme had only succeeded in bringing to village India the false promise of several thousand holes in the ground. Reputations were at stake, not to mention the lives and livelihoods of rural people. Confronted by this debacle, Unicef's new investment was in the balance and those who had earlier been sceptical again raised doubts. For various reasons—political pressure, vested commercial interests and the forcefulness of those convinced of the desirability of pressing on—the upshot was that the show would go on. But no more drilling rigs would be imported until a solution to the handpump problem had been found.

With the urgent pursuit of a better handpump, the first chapter in a different area of technical focus opened (see Chapter 3). Meanwhile, Unicef began to absorb the lessons of the drilling experience to date.

The Halco and Atlas Copco rigs were intended for the hard rock, shallow overburden areas of the country. They had done well. With their small size and low weight, they were very manoeuvrable, reaching many remote rural villages over difficult terrain. But as the programme advanced, more complex hydrogeological conditions were encountered. When drilling in the softer formations in the overburden rather than in places where the hard rock was close to the surface, these small air-operated rigs did not have sufficient rotary power. DTH hammers were not suited to drilling in soil, clay or in the weathered mantle. Where the overburden layer was deep, and inadequate protection provided for the hole, it could cave in under pressure. This was likely to happen where the mantle had become saturated with water during the monsoon. A rig powerful enough to drill a wider-diameter hole was needed to permit the insertion of a casing pipe to prevent the borehole collapsing. There was also the problem of depth. In many areas it was difficult to persuade the smaller rigs to drill deep enough to assure a reliable year-round supply of water. At times when the water table was artificially high due to the rains, the presence of water in the hole increased resistance to the drilling hammer. The resulting boreholes frequently ran dry in the summer.

Developments were underway in the waterwell drilling industry which opened up the possibility of deeper boreholes with large enough diameters to take a casing pipe in the upper reaches of the well. Bits were improving, and new types of hammers driven at faster rates by higher-pressure compressors were now available, markedly increasing drilling speeds. These improvements could be had on rigs which were just as compact and manoeuvrable, by using hydraulics as opposed to pneumatics to operate them. In 1978, Unicef purchased two hydraulically-operated rigs from a UK company, Hands England. There was the usual debate about whether India was competent enough to manage the most up-to-date technology, but once again the advocates for a 'best technology' approach won the day.

The only problem was that, with a large variety of hydraulic rigs now available, Unicef's own engineering professionals were unable to agree on the precise specifications—hole size, potential depth, and configuration of different component parts. A number of purchases were made of very large rigs which had extraordinary capacity for drilling at speed. But they were too unwieldy and over-equipped; few fulfilled the potential their advocates had imagined. There had been a loss of perspective. The idea of 'small is beautiful' for handpump supplies in impoverished village India, where ability to reach remote sites was critical, had been eclipsed. Borehole technology was taking off and Unicef followed unquestioningly. Equipment was selected against capacity and technical criteria alone, not on its all-round suitability for the programme.

A tension had developed, between serving those villages that were most in need—by definition the poorest and most remote—and developing the overall capacity of India's drinking water supply programme. The technological fit, and by implication the social and economic fit, between the programme design and its suitability to meet the needs of the poorest communities, families, and children, in whose name it was primarily justified, had begun to get out of kilter. Once over-mighty rigs had been acquired by PHEDs, the need to keep them occupied and effectively functioning became a more important programme dynamic than the 'problem village' or water scarcity criteria which ought to have been the main determinant in deciding where rigs were to be deployed. What had happened was that the combination of political momentum and considerable financial resources had skewed the programme in certain directions. State governments had taken it up with a vengeance. Maharashtra, for example, acquired a fleet of 60 rigs in 1973–4, and others such as Andhra Pradesh, Karnataka, Madhya Pradesh, and Tamil Nadu, followed suit (Jaitley and Daw 1995).

By 1980, 94,000 borehole handpumps had been installed, theoretically reducing the number of problem villages to around 56,000 (Jaitley and Daw 1995). But a new government survey revealed that there were now 231,000 villages needing water supplies. This vastly increased figure led partly from the decision to once again expand the criteria for problem villages, partly from better surveying, partly from the continually dropping water table, and partly from the inadequacies of some existing wells. But it also stemmed from the political momentum behind the programme, which was influenced by the upcoming IDWSSD declared for 1981–90. The borehole and handpump programme with its strategy of throwing technology at targets was to be India's linchpin response to the Decade challenge of 'Water for All'. The national effort to enhance the rate of progress received a major boost in 1986, with the establishment of the Technology Mission on Drinking Water by Prime Minister Rajiv Gandhi. Its intention was to provide safe drinking water to all the remaining 227,000 'problem villages' by the end of the Seventh Five-Year Plan (1985–90).

Unicef's principal role as an advisor to the Drinking Water Mission was to sort out and streamline the technology. Its thinking ran along the lines that the government laid down. It took as given the idea that village drinking water problems could be solved by what would be described today as a classic supply-driven approach. There was no consultation with villages about what their drinking—or other—water problems were. These were defined by applying the 'problem' criteria. Regrettably, this numbers game itself was easily manipulated by those with a political agenda or some other vested interest. Questions of siting new wells within the designated communities were settled by engineers and drillers, with some help from officials. Occasionally a new borehole could be seen in a district official's or a politician's compound. Little attention was given to issues of land ownership, access for lower castes, or other social parameters which might have emerged from a process of community consultation. There was for the first time an attempt to emphasize the need for health education around water use, and a growing recognition of water quality issues. But these 'software' features were still an adjunct to the main game of borehole and handpump provision—the installation of 'hardware'.

Another key supply-driven feature of the programme was that it offered a 'one-option-fits-all' technological response to the need for safe drinking water—'safe' still being the operable word. The idea that some people might dislike the taste of borehole water and choose not to drink it did not occur to anyone, or, if it did, it was ascribed to their backwardness and ignorance. The 'one-option-fits-all' approach was not only accepted by Unicef, but actively promoted. Groundwater alone was seen as 'safe'. Traditional dug wells were seen as potential sources of disease because they were open and susceptible to pollution. Therefore, however valued they might have been by their users for drinking and other purposes, their contribution to community water supply was ignored.

By the early 1980s, many of the pioneering well-drilling NGOs were beginning to back out. In some cases, this was because their equipment was obsolete. Others, affected by 'appropriate development' ideas then gaining ground, had come to believe that the modern rigs represented just the kind of insensitive high-tech development which helped only the better-off, leaving poor communities even further behind. For example, when new deep boreholes were used for the powered irrigation of water-thirsty cash crops such as sugar cane in Maharashtra, the ponds and dug-wells depended on by local farmers simply dried up (Shiva 2002). It seemed to some of the NGOs which worked close to the ground that borehole programmes had taken on a life of their own and become 'big business', with state PHEDs spending large sums of money on contractors. A few NGO personnel set up Water Development Societies—companies in all but their charitable name—and joined the contractor queue. Others, witness to the unexpected consequences of their enthusiasm for waterwell drilling, now felt that the borehole plus handpump was not the panacea for water problems in poor communities. But within Unicef, the ideology of the handpump–borehole remained sacrosanct.

What the more progressive development thinkers were groping towards was the concept of 'sustainability'. In order for a development approach to bring lasting benefit over the longer term, there had to be a fit between the intervention and the social, economic, and environmental context in which it was applied. Development was not something which could be done *to* people, but only *with* them, in ways that fitted with, and responded to, their reality. However, only after the publication of the Report of the International Commission on Environment and Development in 1987 did the concept of 'sustainability' begin to gain respectability in international thinking. What the Unicef engineering professionals focused on during the 1980s was a more limited concept of 'technological sustainability'. Their main programmatic concern was proving that 'it can be done', and done well—defeating the sceptics over the issue of whether India could operate and manage modern waterwell machinery with all the complications of human resource allocation, supply chains, and ensuring quality installation and repair. They were not unduly concerned by the social fit of the equipment. As for the economic fit, that did not arise since the government poured ever more resources into rural drinking water supplies with every Five-Year Plan. As far as the technological part of the equation was concerned, after a period of confusion in the early 1980s when there was so much sparkling and seductive new machinery around, they did this job extremely well.

An important lesson was that it was better to supply equipment which, in theory, might be in excess of needs, but which, in practice, would ensure that the rig could survive complicated drilling conditions, mistreatment by inadequately trained crews, and the wear and tear of age. However, this did not mean that it made sense to supply monster rigs—the equivalent of a tank when a light armoured car would do. In 1983, Unicef developed a specification for a completely new drilling rig based on experience in all kinds of hard rock and overburden. The key design premise was that a basic rig should be able to accommodate a wide range of drilling conditions using the flexibility provided by hydraulic power. The different components were mounted separately. The mast assembly was in one module; the compressor in a second; the hydraulic power pack in a third. Rigs and compressors were mounted on two small trucks, heralding a 'two-truck' rig for inaccessible rural areas. Different compressors and power packs could be interchanged depending on the requirements at the site and the power capacity required. The manoeuvrability of the earlier rigs was, therefore, maintained—with all the advantages of extra power and flexibility. The configuration of air, hammer, and bit was expected to double previous drilling output—150 boreholes a year instead of 75. Size was kept down. There was no further debate about drilling boreholes for any other purpose than for the installation of handpumps. Power pumps were not Unicef's business, especially as the new handpump, the India Mark II, was beginning to come into its own.

The 'two-truck' rig was to become the standard drilling machine supplied by Unicef from 1984, until the organization phased out of waterwell drilling in 1992. As the technological parameters settled, confidence grew. Unicef made a commitment to support every rig that it purchased with spare parts for 10 years, its anticipated lifespan. Quality norms were also laid down to ensure that all boreholes drilled by Unicef-supplied machinery, and by other rigs purchased or contracted by state PHEDs for their programmes, were drilled deep enough (60 metres, regardless of drilling conditions) to guarantee a year-round water supply, and were cased in the overburden for a minimum of six metres. For a well to be described as successful there had also to be a minimum flow of 12 litres per minute—enough to feed an India Mark II handpump. These 'minimum quality standards' were introduced in 1984 and incorporated in guidelines issued by the GOI for all state drilling operations in the hard rock areas of the country. They were instrumental in establishing a higher and more consistent quality of waterwell than before.

With the introduction of the new drilling norms came the need to monitor rig performance more closely. The rapidly increasing number of wells being drilled prompted the design of a new, computerized rig-monitoring system to ensure maximum output and minimum 'down-time' of each rig. A management tool for planners, engineers, and managers, it not only accommodated the new minimum well drilling standards of depth, casing, and yield, but it introduced the concept of measuring performance against a set target of borehole metres drilled, according to the local setting and the type of rig. Significantly, it monitored the distance travelled by each rig so that the distance between sites could be checked following the logic that rig travel time is lost drilling time. The analysis of the data produced by the system confirmed the irrelevance of investing in higher-capacity drilling machines. A rig of twice the capacity did not achieve twice the output.

Checks and controls to maintain the quality of drilling for rural water supplies were practicable so long as the programme was entirely government-run. But during the 20 years of its existence the programme had radically changed. Far from proving unable to manage drilling equipment, the private commercial sector in India had taken to it like a duck to water. Long before the marketplace ideologues of the 1980s and 1990s began to advocate partnerships between government and private sector, the waterwell drilling business in India provided a copybook example. Technology transfer came about in less than a decade—with its disadvantages as well as advantages. Partly by intention, partly because they were pressing at an open door, Unicef was its agent. It would turn out to be a legacy of which to be both wary and justifiably proud.

The development of an indigenous Indian waterwell industry was initially prompted, as was everything else, by the NGOs. In 1968, at the height of a drought in Andhra Pradesh, the Water Resources Unit of a Methodist Mission in Hyderabad, run by an Englishman, Peter Wood, enlisted AFPRO's help in procuring two Halco Tiger rigs for a crash programme to substitute boreholes for dried-up open wells. Almost immediately, Wood ran into problems with breakdowns and spares. He, therefore, attempted, with some success, to manufacture small replacement parts. Before long, production expanded into hammers and bits and, in 1972, the Water Development Society (WDS), by now an independent registered charity, built its first three rigs (Black 1990). This was the first independent local venture in drilling rig production. Although WDS was unique—its sole mission was to expand water supplies in poor communities, with drilling rig production as a means to this end—it spawned many others which were more commercially, and less idealistically, inspired.

When Unicef embarked on its programme, it was a cardinal principle that companies from which rigs were imported should have local Indian subsidiaries and the capacity to support imported equipment with service and training. Thus, the concept of 'indigenization' was mooted from the outset, and was seen as critical to the task Unicef had set itself—long-term technological sustainability. It was anticipated that spares, and eventually rigs themselves, would be locally manufactured. Besides the practical need, there was also an ideological dimension. At this time, the commitment of India's post-Independence leaders to developing India's own industrial base by a similar route to the centralized, socialistic strategy earlier pursued by the USSR (the erstwhile Union of Soviet Socialist Republics) was firmly in the ascendant. Prime Minister Nehru, like many developing country nationalists of his day, deeply mistrusted what he saw as the imperialist tendencies of international capitalism (Tharoor 2000). These were the forces he held responsible for the suffering of millions of his countrymen under the regime of resource exploitation and wealth extraction represented by British colonial rule. In the first decades of independence in India, international companies operated locally only on terms controlled by the national drive for import substitution. Nehru's dictum had been that self-sufficiency was paramount. 'Be Indian, buy Indian' became a mantra, encouraging the replacement of the inferiority complex of the colonial era with national pride and self-esteem.

Those committed to an Indian vision of development, which included many in the NGO community and in international organizations, had every sympathy for Nehru's vision. They included the original advocates for Unicef's involvement in waterwell drilling. Their conviction that India was able to handle the sophistication of the equipment stemmed partly from their psychological distance from the old colonial mentality. But even they could not have predicted the speed with which local manufacture of hammers, bits, and eventually drilling rigs, would take off. WDS in Hyderabad, which was manufacturing 100 rigs a year by 1982, was only one example. Some of its own engineers soon saw the possibilities of rig manufacture and set them up by themselves. These local entrepreneurs were highly inventive and quick to take advantage of technical improvements introduced by the larger companies.

As the programme grew dramatically during the 1980s, the possibilities for the local manufacturer and contractor of getting a slice of the action became increasingly apparent. The demand was enormous, as was the political backing. Whereas in 1974, GOI outlay for the rural water supply programme was around US$ 20 million, by 1989 it had risen to US$ 600 million (Jaitley and Daw 1995). The existence of a very large programme with significant government resources acted as a powerful incentive to contractors on the one hand, and local engineering departments on the other. The Unicef formula—sound equipment from the best companies, local back-up, feedback from rig operators to the manufacturers—provided the building block for technology transfer. Unicef accepted that indigenization did require trade-offs: the large companies produced higher-class and more reliable components and stuck more closely to delivery schedules and specifications. But compromise, despite reduction in product quality, seemed worthwhile. Unicef adhered strictly to import substitution as its policy, and thus helped promote an extraordinarily successful, locally-driven transformation of external technology into a thriving local manufacturing industry.

The turning point came with the decision to go down the hydraulic route. The small-scale rig industry concluded very quickly that if Atlas Copco, Ingersoll Rand, and Halco could make rigs using outsourced hydraulic components, why not everyone else? The introduction of hydraulics with their high degree of reliability spelled the beginning of the end of their part in waterwell equipment manufacture for many of the large, established companies in India. The small-scale manufacturers could build rig frames and masts cheaply and mount identical hydraulic components and also guarantee identical reliability. They were able to undercut the prices quoted by the international companies by around a third. And over the next 10 years that is exactly what they proceeded to do.

By the late 1980s there was a mushroom growth, particularly in south India, of do-it-yourself engineers copying imported drilling equipment in back alleys and garage workshops. The dry time of year became a well-established six-to-eight-month 'drilling season' when the water table dropped and demand for well drilling increased.

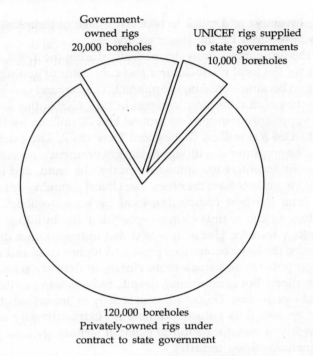

FIGURE 2.1: Boreholes drilled by different agencies: 1992–3

Sources: GOI, state governments, and Unicef Rig Monitoring Systems.

From around October to June, when the summer heat took hold and water sources dried up, a small army of drilling entrepreneurs would set off for states such as Karnataka, Madhya Pradesh, and Rajasthan to sell their services for the duration. When the rains came, they would gravitate back home. Many were based in Tiruchengodu, near Salem, in Tamil Nadu. Over the monsoon months, hundreds of rigs would be parked in the streets throughout the town, undergoing repairs in preparation for the next drilling season.

By the late 1980s the writing was clearly on the wall as far as state PHED drilling operations were concerned. As government rigs aged, their borehole production dropped and operating costs climbed. Meanwhile, the equipment of the local contractor improved in quality and his drilling costs plummeted. With government drilling capacity on the wane, competition between contractors intensified as they vied for business from PHEDs. These were scrambling to meet their water supply targets in the face of the political demand for still faster rates of coverage. The contractors came into their own in the severe drought of 1987. They were able to respond quickly and efficiently to crash programmes of borehole drilling, and often managed an output three times higher than that of ponderous government operations.

However, the shift into 'public–private partnership' brought some serious disadvantages. The quality norms of minimum depth, minimum casing, and minimum yield for drilled wells, so carefully introduced with Unicef assistance, became impossible to monitor and enforce. State PHEDs did not have the means to visit each borehole site to keep pace with a contractor drilling round the clock, sometimes managing an output of three wells a day. Monitoring became the task of the contractor himself, who was paid according to his own testimony of compliance for each well he drilled. This arrangement was open to abuse, although communities were becoming increasingly familiar with well-drilling techniques and were wise to the wiles of the unscrupulous contractor. They counted the drill pipes as they came out of the ground and so were seldom shortchanged—at least on drilling depth.

But there was a downside to indigenization of the waterwell industry. Once drilling became 'contractor oriented', competition within the programme between the profit motive and the public service motive jeopardized quality standards. Outside the programme, the business of borehole installation began to cater to a mass market of farmers and better-off householders. Before long, this was to propel the issue of environmental sustainability to the point of crisis.

In 1987, India suffered the worst drought of the century. A state of emergency was declared in several states, including Madhya Pradesh, Maharashtra, Gujarat, Orissa, and Rajasthan. In Rajasthan, livestock camps were established and long lines of bulging fodder trucks plied the roads daily from Haryana and further south. In Gujarat, trains were commandeered to haul water, and in towns such as Rajkot in Saurashtra, water was distributed to houses by tractors and tank-mounted rickshaws. In Maharashtra and Andhra Pradesh, tankers distributed water to villages every few days. So serious was the crisis that at one stage there were fears that it might overwhelm the Union government.

India pulled through. And the crash programmes of drilling and well-deepening needed to save lives and livelihoods convincingly demonstrated the maturity of the national waterwell drilling industry and the capacity of contractors and state PHED programmes. The effectiveness of the relief programme, in which Unicef played a part in all the worst-hit states, was rightly a matter of national pride. But there had been a scare. The age-old scourge of famine had nearly raised its head. The two consecutive years of drought and the apparent volatility and capriciousness of the monsoon were a grave cause for concern. The National Drinking Water Mission set about taking a hard look at rural drinking water supplies in the country. One of its 55 mini-missions was devoted to water resource sustainability.

The increasing availability of more sophisticated drilling technology to identify and reach ever deeper groundwater aquifers had had an unfortunate by-product—higher levels of groundwater extraction than the resource could tolerate. This was especially the case in the vulnerable hard rock areas with their inherently limited storage and slow replenishment rates. Aquifers are not troubled by a low-output handpump installed on a slim borewell unusable by a powerful motor-driven pump. But the uncontrolled use of such pumps on larger diameter boreholes was drawing down the water table. Handpumps in the vicinity of these voracious wells were more frequently becoming defunct, putting drinking supplies in jeopardy. The progressive drop in groundwater levels coupled with the rapaciousness of the drilling industry to satisfy the market for irrigation boreholes were harbingers of an emerging crisis.

In 1987, the government issued a new national water policy, declaring that: 'Water is a prime natural resource, a basic human need and a precious national asset.' (Nigam et al. 1998) Anxiety about the threat to the country's groundwater was clearly expressed: 'Exploitation of groundwater resources should be regulated so as not to exceed recharging possibilities...'. The drinking water needs of both human beings and animals 'should be the first charge of any available water.' Unfortunately, nothing could illustrate better the gap that can occur between high-sounding policy statements and the lack of a practical and legal regime to make possible their implementation. Whatever the concerns expressed about the implications of ground-water over-extraction for future water supplies, existing farming and energy policies stood in the way (see Chapter 7).

While Unicef was reluctant, for understandable reasons, to address such a complex skein of interests, there was a technological contribution it could make to reducing the impact of water table decline on drinking water supplies. During the 1987 drought, the National Drinking Water Mission approached Unicef with a request to provide hydrofracturing units for revitalizing existing boreholes (UNICEF 2000). Borrowed from the oil industry and extensively used in Scandinavia and the US, this technique pumps water at very high pressure into a dry or low-yielding borehole. Where a borehole's output has declined, the reason may be that silt has accumulated in the fractures, obstructing the flow of water into the well. Hydrofracturing is a technique for forcing open the network of fractures supplying the borehole and flushing them free of sediment, thereby stimulating their flow. In certain formations, results may be even more dramatic. The water pressure from the hydrofracturing process is powerful enough to split the rock, thereby increasing the network of water-bearing fractures serving the borehole.

Villagers in places where the water table has dropped significantly are in a constant state of anxiety year upon year at the prospect of wells running dry. As the dry season advances, they may be forced to resort to sources located many miles away, or which are unprotected, stagnant, and clearly unsafe. Countrywide, an estimated 50,000 wells are abandoned every year because their yield has become inadequate. A rule of thumb estimate is that around half of these can be technically restored, either by hydrofracturing or by flushing with compressed air. Thus, for substantial numbers of communities, well rejuvenation is a salvation. Hydrofracturing can also be used to improve water yields in cases where the initial well is unsuccessful. A typical new borehole failure rate is between 10 per cent and 15 per cent (UNICEF 2001).

Unicef imported the first eight hydrofracturing units in 1989 and deployed them in Madhya Pradesh, Maharashtra, Gujarat, and Rajasthan. They performed well, achieving a 70 per cent success rate.

Some boreholes were able to deliver many times their original yield. In 1998, the National Drinking Water Mission calculated that around 200,000 marginal boreholes were suitable for the application of hydrofracturing technology. However, the use of the technique is only a short-term solution. It makes better use of the available aquifer rather than create new sources. Hydrofracturing does nothing to reverse over-extraction. It simply delays the inevitable. Recharge is still essential. Nevertheless, hydrofracturing demonstrates that, for about one-third the cost of a new well, an old well can be rejuvenated and even vastly improved, thereby reducing the investment costs in new wells and limiting their number.

For several years, Unicef supported hydrofracturing in many drought-affected states, importing equipment to demonstrate the principle of hydrofracturing and encourage the use of hydrofracturing units by PHEDs and contractors to rejuvenate old wells. By 1998, the national fleet comprised 40 units (UNICEF 2000). Madhya Pradesh and Maharashtra both achieved impressive results. But despite Unicef's best efforts, it did not manage to persuade contractors to invest in this equipment. Hydrofracturing does not give the same financial returns as drilling. Only when the government and the PHEDs become sufficiently exercised about environmental protection and the effects of uncontrolled and ever-deeper borehole drilling will they force the pace with the commercial sector. With the poor rains of 2002, state governments were entreating Unicef for hydrofracturing assistance. But by then, Unicef's drilling equipment days were over.

Unicef's last purchase of drilling rigs was in 1992. The remaining drilling niche it filled, from 1988, was to provide rigs to reach places inaccessible even to the small, two-truck mounted rigs. An all-terrain vehicle was needed, of a type unavailable in India. So, exceptionally, Unicef imported special rig carriers. With their all-wheel drive and front and rear axle steering, these machines could clamber over anything. Halco constructed a highly successful upgraded Tiger using these carriers. But Unicef made it clear that its purchase of such machines would be restricted to a few 'special needs' areas and there would be no attempt on its part to indigenize the technology. Ingersoll Rand also managed to purchase some extra-rugged, all-terrain vehicles specially manufactured for the Indian Army and normally unavailable for commercial use, on which to mount their rig. Theirs was the last kind of drilling machine Unicef bought.

Spare parts and back-up were provided for the agreed 10-year period, after which, finally, the remaining Unicef operations with regard to drilling were wound down. Unicef had supplied, over the 25 years between 1967 and 1992, 330 drilling rigs, at a total cost (at the time of purchase) of some US$ 33 million (UNICEF 2000). By this time, India was fully equipped to manage its rural drinking water supply programme from its own financial and technological resources. Within Unicef, there was a strong feeling that the time had come to concentrate its own efforts almost entirely on software: piloting ways to change attitudes and hygiene behaviour—as it had previously piloted technology—so that the health benefits of water supply and sanitation inputs could be realized.

In the run-up to the first Earth Summit in Rio de Janeiro in 1992, there was a growing sense of urgency worldwide about environmental issues. For the first time, water was being spoken of as a finite and precious resource with an economic price-tag. Both in India and internationally, 'sustainability' had become the new watchword. With understanding about the finite nature of the resources came growing awareness that there could be no sustainability of water services unless communities were themselves directly involved in running and maintaining community installations such as wells, boreholes, and handpumps and contributing to their costs. These ideas were, belatedly, also penetrating the Indian rural drinking water supply programme. For long regarded as one of the most progressive and successful rural drinking water programmes in the entire developing world, the supply-driven service model of which it was the archetypal exponent was now beginning to look somewhat old-fashioned.

And sustainability of the resource itself? What was to be done about that? Nobody yet seemed to know or, to put it more appropriately, to care.

References

Black, Maggie (1990), *From Handpumps to Health*, UNICEF, New York.
————— (1987), *The Children and the Nations*, UNICEF, New York.
Dey, A. K. (1968), *Geology of India*, National Book Trust, India, New Delhi.

Jaitley, Ashok and Raj Kumar Daw (1995), *Contribution of Voluntary Organisations in Rural Drinking Water Supply and Sanitation Programmes in India*, paper sponsored by UNICEF, Delhi and presented to the Water Supply and Sanitation Collaborative Council in Barbados, UNICEF, Delhi, unpublished.

Nigam, Ashok, Biksham Gujja, Jayanta Bandyopadhyay, and Rupert Talbot (1998), *Freshwater for India's Children and Nature*, UNICEF and WWF, Delhi.

Shiva, Vandana (2002), *Water Wars: Privatization, Pollution and Profit*, Pluto Press, London.

Tharoor, Shashi (2000), *India from Midnight to the Millennium*, revised edition, Viking/Penguin India.

UNICEF (2001), *Well Rejuvenation*, Field Note, UNICEF India Country Office.

————— (2000), *Learning from Experience: Evaluation of UNICEF's Water and Environmental Sanitation Programme in India, 1966–98*, Evaluation Office, UNICEF, New York.

3
The Handpump Revolution

In 1974, just as the new approach to the provision of rural drinking water supplies was getting into its stride, a discovery threatened to ruin the entire programme. As noted in Chapter 2, surveys had been carried out by Unicef in Tamil Nadu and Maharashtra which showed that 75 per cent of the handpumps installed on the new boreholes were out of action. The technical problems to do with drilling had eclipsed other issues. A handpump, after all, was a well known and unsophisticated device. The teams of drillers with their high-powered equipment had gone to the villages, performed their marvels, installed the pumps, and disappeared. The focus of technical attention had all been on the borehole; but the simple end of the technology had turned out to be the one to cause disaster.

The problem was two-fold. In the first place, the cast-iron pumps used were poor quality copies of old-style European and American handpumps, long since out of use in the land of their birth except as ornaments on garden water features. Designed to be used by a single family, they were not up to the wear and tear of use by a community of over 500 people for around 10 hours a day, and they broke down repeatedly. In the second place, false assumptions had been made about the capacity of communities to maintain them. The *gram panchayats* (village councils of elected representatives) were expected to take them into village 'ownership' and have them fixed up when necessary by local *mistris*—handymen operating bicycle and water pump repair shops. But no system of spare parts provision had been made, nor had the mistris been trained in handpump maintenance.

It was not clear to the villagers what to do about the breakdowns, nor to whom, if anyone, to report them. So they simply waited for the miracle men who had put the pumps there in the first place to reappear and mend them, and resorted to the wells and other open sources they had relied on in the past.

Unicef was on the point of making a new US$ 9.3 million investment in the programme, and the discovery that three-quarters of the beneficiaries had, up to now, gained little from it but a hole in the ground was discouraging to say the least. Fortunately, mission-run water schemes in Maharashtra had already faced the same problem and had come up with some new, tougher handpumps. One of these was a pump developed in the late 1960s at the Church of Scotland mission at Jalna—the same mission that had pioneered high-speed drilling under John MacLeod. This pump was made of steel, with a single pivot handle and a sealed ball-bearing (Mugdal 1997). The life of the connecting rod was improved by being kept aligned and in a state of constant tension by means of a link chain running over a quadrant. Several hundreds of these distinctive yellow Jalna pumps were manufactured and installed on local wells.

In the early 1970s, an advance on the Jalna was made by Swedish engineers Eric Jallen and Oscar Carlsson based at a mission in Sholapur, Maharashtra. The pivot mechanism was improved, the link chain replaced with a roller-chain, and the pedestal was designed to fit neatly over the borehole casing pipe. Set in concrete, it provided a sanitary seal so that potentially contaminated material could not fall back down the hole. The Sholapur pump, which used jigs and fixtures so that the parts were uniform, was also more professionally manufactured than the Jalna one. Other adaptations of these pumps began to appear in different parts of Maharashtra and were occasionally exported elsewhere. But, overall, these developments were haphazard, with little thought for standardization. Guaranteed interchangeability of handpump parts was, therefore, impossible.

In the 1974 crisis faced by the national programme, Unicef's water engineers turned to the Sholapur pump as a short-term solution, and purchased 6500 conversion heads to allow state governments to fit onto the cast-iron pedestals. These pumps did much better than those they replaced, convincing the authorities that an altogether stronger and better-engineered pump was needed for the rural water supply

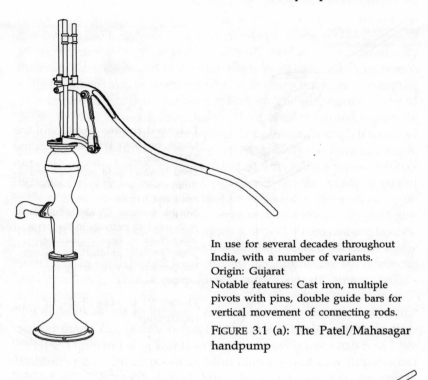

In use for several decades throughout India, with a number of variants.
Origin: Gujarat
Notable features: Cast iron, multiple pivots with pins, double guide bars for vertical movement of connecting rods.

FIGURE 3.1 (a): The Patel/Mahasagar handpump

Developed between 1969–74.
Origin: Church of Scotland Misson, Jalna, Maharashtra.
Later modified by the American Marathi Mission, Vadala and called the Jal–Vad handpump.
Notable features: Fabricated steel body, quadrant handle with link chain and sealed ball bearing pivot.

FIGURE 3.1 (b): The Jalna handpump

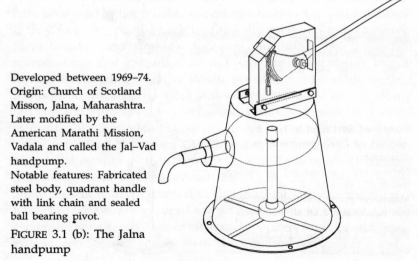

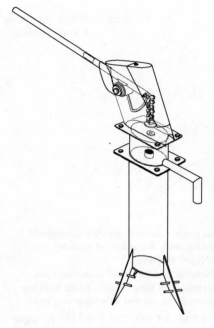

Developed between 1969–75.
Origin: Sholapur Well Service,
Maharashtra.
Used standard mild steel pipe and
fittings with a mild steel, fabricated
head and handle.
Notable features: Quadrant handle
with two ball bearings in the pivot,
roller chain on handle quadrant,
pump pedestal groutable in concrete
independent of well casing, for
sanitary seal.

FIGURE 3.1 (c): The Sholapur
handpump

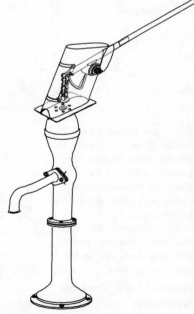

Developed for Unicef in 1974, by
AFARM, an NGO consortium in
Maharashtra.
Notable features: Adapted the Sholapur
handpump head to the Patel/
Mahasagar pump body, eliminating cast
iron handle and multiple pin pivots.

FIGURE 3.1 (d): The AFARM
conversion head

programme, especially in hard rock areas. Various existing options were considered and rejected, and from 1975 Unicef found itself taking the lead in the development of a new locally-manufactured Indian handpump. In default of anyone else to take up this challenge on a countrywide scale, Unicef fell into the role of coordinator and facilitator. The eventual outcome was the India Mark II.

Unicef's key partners were the government's Mechanical Engineering and Research Organization, a part of the Council for Scientific and Industrial Research (CSIR), which offered design support, and Richardson & Cruddas, a government-owned engineering company in Chennai, which manufactured the prototypes, tested them, and rationalized components with a view to mass production. The Sholapur pump was the starting point, although it was modified in various ways. The key parameters for the new pump were that it could be locally produced using components and materials available in India, that it would be easy to operate, would be sturdy and able to function without breaking down for at least a year, and that it should be able to draw water from borewells whose water level was at least 30 metres below the surface. It should also be inexpensive—less than US$ 200 at current prices.

Although Unicef led the handpump development project, Richardson & Cruddas bore the research costs without any guarantee, on the expectation that a successful pump would garner significant numbers of state contracts to make it a commercial proposition. Field-testing of 12 prototypes was carried out in Coimbatore, a district of Tamil Nadu where the water table was low (around 40 metres) and handpump usage high. When, in October 1977, the pumps completed their one-year intensive test with only one failure, the future of the India Mark II looked assured. A further 1000 pumps were produced and put on trial in various states. Their low breakdown rate was confirmed, they were easier to operate than the old cast-iron models, and communities liked them. State governments began to get interested.

Production of the India Mark II began in earnest in 1978. Richardson & Cruddas manufactured around 600 pumps a month through a network of small-scale ancillary workshops employing around 800 workers, to produce pump parts. Jigs and fixtures were supplied and technical assistance given to establish mass production facilities.

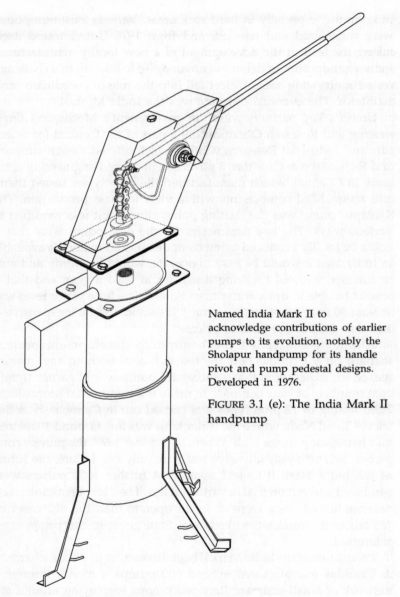

Named India Mark II to
acknowledge contributions of earlier
pumps to its evolution, notably the
Sholapur handpump for its handle
pivot and pump pedestal designs.
Developed in 1976.

FIGURE 3.1 (e): The India Mark II
handpump

FIGURE 3.1: Evolution of handpumps in India
Source: All drawings by Raj Kumar Daw.

Unicef put considerable energy into advocating the use of the India Mark II in hard rock areas, and was itself one of the earliest large customers for the pump, using it for a programme of 'rejuvenating' wells whose useless pumps had rendered them defunct. It also invested in training for PHED staff in the pump's installation and maintenance procedures. The performance of the distinctive-looking pump turned out to be its own best advertisement. Although there was some initial resistance from conservative state engineering departments, the benefits of a pump that would lower the breakdown rate and raise the quality of the programme by a significant margin quickly overcame all scepticism. Before long, 10 states were fully converted to the new pump—Andhra Pradesh, Bihar, Gujarat, Karnataka, Madhya Pradesh, Maharashtra, Orissa, Rajasthan, Tamil Nadu, and West Bengal (Black 1990).

Efforts began at an early stage to develop manufacturing capacity in other parts of India. It was deliberately decided not to patent the new handpump but put it into the public domain so that wide-scale production could open up as fast as possible. The designers of the Sholapur pump had reached the same conclusion. However, the development of the new device for which manufacturing capacity in India readily existed, and for which there was a sizeable market shaping up, was fraught with risk. Its very success in becoming a standard item of procurement in many state rural drinking water programmes was a potential threat in the Indian context. Unless quality control could be assured, cheap and shoddy imitations of the India Mark II would flood the market. This would not only make the standardization of parts impossible, but threaten the fundamental precept of the pump. It would cease to be reliable, be subject to repeated breakdown the technology would be discredited, and the whole programme would be back to the point where it had faced collapse in 1974.

For this reason, Unicef was anxious to assume a guiding—even a controlling—role in the development of the India Mark II handpump industry. It was, as in the case of borehole drilling, anxious to protect its investment and to see that its purpose—better access to water and the provision of health care for mothers and children—was not compromised. Today, when 'public–private partnerships' and devolution to the private sector of responsibilities previously assumed by the public sector are the flavour of most development policies, the lessons of the Mark II experience are as pertinent as those connected to waterwell drilling. Above all, they indicate the pitfalls of working in an environment in which regulation is honoured better in the breach than in the adherence, as well as the need to 'stay with' a product to assure its quality in settings where the ultimate consumers have neither the knowledge nor the leverage to guarantee it for themselves.

In 1979, a national conference on deep-well handpumps was held in Madurai, Tamil Nadu under the auspices of the CPHEEO and Unicef. Representatives from seven states attended, and the conference recommendations moved the rural drinking water supply programme throughout the country several steps further towards uniformity and standardization on the India Mark II as *the* programme handpump. The conference fully endorsed the need for the pre-qualification of suppliers of the handpump so as to assure interchangeability of spare parts and resolved that all states establish quality control measures. But it was one thing to agree to such a process, quite another to see it implemented and enforced. Manufacturers were already turning out all kinds of versions of the handpump which states with lax quality control were happily installing. A better route to standardization and quality assurance had to be found.

In 1980, the Indian Standards Institute (later the Bureau of Indian Standards (BIS) issued the first full specification for the pump's manufacture: IS: 9301–1979 (UNICEF 2000). At the same time, Unicef engaged the Crown Agents—originally a British government body but now an international inspection agency—to carry out works inspections and 'clear', or otherwise, potential manufacturers. The Crown Agents also provided technical assistance to enable manufacturers to iron out production and quality-related problems. A number of potential manufacturers came forward, including Inalsa, Meera and Ceiko Pumps, and Ajay Industrial Corporation. These and others managed to establish effective production and in-house quality control systems for pump parts, tools, and assembly. Only when their plants had been inspected, and they had demonstrated that they could produce pumps to the desired quality, could they join the

Unicef list of qualified India Mark II suppliers. State programmes bought only from those companies on the list. The degree of Unicef control over quality conferred by the government was a unique feature of the programme.

By 1984, Unicef had approved licence to manufacture to 36 companies with an annual production capacity of an extraordinary output—200,000 pumps. The combination of standardization, quality control, and incentive to private industry paid rich dividends. Unicef, the government, and the manufacturers continued to work on technology refinement through research and development (R&D) and field testing to improve reliability, durability, and ease of maintenance. New revised standards were issued in 1982, 1984, and 1990, as various improvements in design, material, and components were tried and tested.

Unicef's role in the development and promotion of the India Mark II pump was already, therefore, rapidly evolving. By the early 1980s, it had ceased to be a major purchaser of pumps itself, only buying around 6000 a year for the tail-end of the 'rejuvenation' programme to replace cast-iron pumps. Instead, it became the financier for the system of manufacturer approval, and for pre-delivery quality assurance inspections of all handpumps and spare parts leaving the factory. This meant that every single pump bought by state governments for their drinking water supply programmes, and every spare part, was checked before it went off to be installed, courtesy of Unicef. This did not inhibit or hold up production. By 1984, more than 600,000 pumps had been produced and installed (Mudgal 1997). Between 1985 and 1987, 430,000 pumps underwent Crown Agents' inspection.

So important did Unicef regard the need for quality assurance for handpumps that it continued to pay for this service until 1996, by which time 62 handpump manufacturers were licensed with an annual production output of 300,000 pumps, some of which were being exported. For another two years, Unicef continued to pay for spare parts inspection, finally releasing its role in quality assurance fully to the BIS in 1998.

These twin emphases—standardization and third party quality control—and the role of Unicef in ensuring them, were cornerstones of the India Mark II handpump experience, and were key to the pump's adoption throughout the country. Unicef managed to act as a bridge between the need for a particular type of water-lifting device at the local level, the potential Indian manufacturers of that device, and the large-scale institutional and logistical requirements of government service delivery. Not only were pump parts interchangeable from one end of the country to the other, procurement was simple, as was inventory control. Very few basic tools were needed for its repair, and these too were standardized. Availability was easy, and competition among manufacturers helped keep India Mark II prices down. All in all, the potential economies of scale for a countrywide programme of water supply service delivery were realized in a careful and professional manner.

The India Mark II rapidly became a household name, its reputation assured, and its presence in the landscape ubiquitous. Undoubtedly, long-term commitment to handpump quality by Unicef was important in making this happen. This form of technical support gave time for companies to develop the manufacture of the pump to a consistent and high standard—which also helped gain significant orders from abroad. From the early 1980s, when the India Mark II was given a seal of high international approval by a UNDP/World Bank handpump testing programme, the pump began to be exported to countries in Africa and Latin America—initially for use in other Unicef-assisted programmes. The strong emphasis on standards and inspection also taught state procurers and implementing agencies to appreciate the need to maintain the quality of both the pump and the spare parts, and to develop their own capacity for quality control.

Any external support organization has difficulty deciding at what moment to withdraw from a programme, or to cease fostering one of its linchpin components. By the mid-1990s, when well over 2 million India Mark II handpumps had been installed, there was a strong feeling in Unicef that it was time to stand back from the technological side of handpump water supply provision. Its key technological contributions—hard rock drilling and the introduction of a standardized deep-well handpump for low water table areas—were now fully absorbed into India's own industrial economy. There was long experience with both aspects of the handpump–borehole technology, and all the technical skills and managerial competences were now well-established in state implementing bodies. Private expertise, should it be needed, also existed in every area. Dependency on an external partner—a pitfall in any venture in international cooperation—would also be avoided with the timely withdrawal of Unicef from the programme. By this stage too, the organization had shifted its own focus of attention in water supply and sanitation, and become more interested in the 'software' elements: social acceptability, usage, health awareness and its impact, personal hygiene, and sanitary behaviour. Concern with 'hardware' was definitely out of fashion with Unicef.

In 1996, the responsibility for quality control was transferred to the BIS. The list of qualified manufacturers is maintained by BIS, who also conduct a pre-delivery inspection at factories, for which the state departments pay. But no one pretends that the quality of the quality control is what it used to be. Some manufacturers believe that BIS has neither the capacity nor the management capability to ensure unbiased inspection in the same way that independent third-party inspections guaranteed (UNICEF 2000). One reputable company, Inalsa, has ceased to manufacture India Mark II pumps for the local market because prices have been forced down to the point that corners on quality have to be cut. Because inspection is not conducted on every pump and is not adequately stringent, cheap and shoddy pumps are able to compete favourably for contracts with the properly-made ones. The potential profit margin on a pump is thereby rendered unattractive, even unfeasible, to a reputable manufacturer.

The issue is an awkward one. It is difficult to argue that Unicef should have paid for Crown Agents' inspection indefinitely. After 17 years, it was logical for it to be passed on to the states'. It ought to have been possible in that time to embed the principle of quality control into the handpump programme and transfer the role to state and national institutions. If it was not possible, how long would it take? In the relationship between the supply and demand for a service whose efficacy depends on technical high standards, there have to be sanctions or incentives to ensure the maintenance of quality. In India, the sanctions and incentives too often operate in the wrong direction. Mutual self-interest between purchasers (state bureaucracies) and suppliers (contractors) is not yet adequately moderated by the expression of the interests of service users, operating through democratic or consumer mechanisms. In retrospect, more could have been done earlier in the programme to ensure that the representatives (village leaders and local officials) of the service users could recognize the difference between a shoddy pump and a good quality pump, and reject the former. Instead, all responsibility for maintaining the quality of hundreds of thousands of handpumps manufactured annually in different parts of the country was handed to a centralized agency, the BIS. This was short-sighted.

This issue interacts with another. The triumphant success of the Mark II plus borehole for rural water supplies eclipsed all other solutions, at least in the hard rock areas, and in areas with hydro-geological formations where the rock was less hard and the water table less deep. In some parts of the country—in West Bengal and in Uttar Pradesh, for example—the Mark II was installed in areas where there was no need for a pump of this expense and sophistication because the water table was relatively high. The argument was that it was sturdier and more sanitary. But by using this technology, the management of water supplies was effectively removed from the people who had previously depended on a simpler type of pump and tubewell, which local panchayats could afford to install and main-tain. The changeover, therefore, suited bureaucrats, politicians, and contractors but did little for service users. Where the pump was poorly installed, and water collected in cesspools around its base, the environment and the water supply was not more sanitary but less so. When the pump broke down, they were dependent on outsiders to repair it.

Since the 1970s, the supreme value attached to groundwater as the desirable source for domestic water supplies—because it is assumed to be 'safe'—meant that open sources have been definitively regarded as 'unsafe'. Yet in some places, its safety, even at the source, was compromised by poor installation. In others, open sources were and still are preferred for reasons of taste, and if the water is filtered or its turbidity reduced, there is no reason why it cannot be made 'safe'. Where local people do not wish to consume handpump water, they may have science on their side: it may be brackish, or contain some other chemical contaminant such as iron.

In 1984, a survey was conducted on handpump use in four dis-tricts in different parts of the country: Ajmer in Rajasthan, Jhabua in Madhya Pradesh, Mayurbhanj in Orissa, and Tirunelveli in Tamil Nadu (UNICEF 1985). The study showed that, where no suitable alternative existed, people did rely heavily on handpumps, and that they were especially important for the poorer villagers. But it was also discovered that only 60 per cent of those with access to handpumps used them. Although people were prepared to walk long distances to wells, they would not walk more than 150 metres to a handpump. Beyond this distance, usage rates were low, and beyond 300 metres, handpumps were not used at all. Location was one significant reason. Women fetching water did not like public places such as the village panchayat office, the bus stand, school or marketplace. People also tended to believe that handpump water was inferior. They did not appreciate its quality of 'safe' and thought that rice cooked in water from traditional open wells tasted better.

The importance of the taste of drinking water was borne out by another survey commissioned in 1988–9 on people's views and behaviour around water, including which type of sources they preferred for what type of purposes (UNICEF 1990). In spite of nearly 20 years of commitment to the unique 'safety' of handpump water for drinking, it turned out that—except in Uttar Pradesh and West Bengal where for many years the people had been using suction pumps fitted on tubewells installed by traditional methods—the open dug well continued to be the people's chosen source. In all other states, 13–18 per cent of the people never drank handpump water. Half the respondents (of whom there were 7900 in 22 districts, four out of five of whom had a handpump in their village) said the handpump was too far away. One third said its water tasted nasty, looked rusty, or smelled medicinal. These findings came as a shock. It had been assumed the handpump had become the primary source of water for drinking, and that its use for this and other household purposes was considerably higher than was in fact the case.

In some places, other dynamics govern consumer desires. In Rajasthan, when offered a handpump instead of an open well, many communities reject it because six people can draw water from an open well simultaneously, whereas only one can operate the handle of a pump. Queuing may absorb hours of time which rural women can ill-afford. One study of the utilization patterns of handpumps in Orissa supplied by a four-year programme between 1988 and 1992 tried to establish why a high proportion of borewells had fallen into disuse. Some handpumps yielding water that was non-potable according to chemical tests were being used for drinking and other

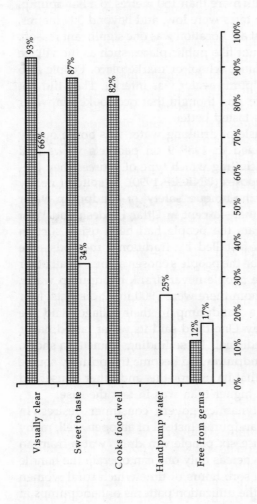

FIGURE 3.2(a): Attributes of 'good water'

Drinking Water: Users' and programme implementors' perceptions

□ Implementor's view   ▦ User's view

Visually clear — 93% / 66%

Sweet to taste — 87% / 34%

Cooks food well — 82%

Handpump water — 25%

Free from germs — 12% / 17%

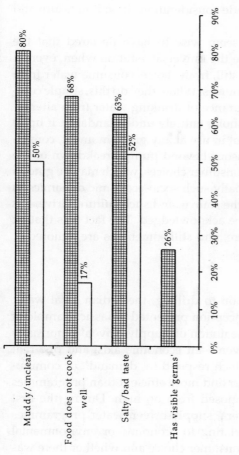

FIGURE 3.2(b): Attributes of 'bad water'

Drinking Water: Users' and programme implementors' perceptions

Source: Water, Environment, and Sanitation: A Knowledge, Attitudes, and Practices Study in Rural India, 1988–9, by Indian Market Research Bureau, for Unicef.

□ Implementor's view   ⊞ User's view

purposes; others that were technically safe were rejected. After analysis, data on 4000 wells revealed that the only clear-cut reason why people used a handpump supply as compared to any other was if no other source was available or better located; not mere 'convenience' but economic necessity overrode considerations based on health and taste (Daw 2002).

In retrospect, it does not seem wise to have declared that the handpump–borehole has to be the universal solution when, even in some hard rock areas, people still, from choice, consume water from open wells and other surface sources. Where they do this, simple comparison of the taste and appearance of drinking water from alternative sources may make their choice entirely understandable. If up to 40 per cent of borewells are not in use at any given moment, consideration has to be given to reasons beyond pump breakdown or the lowering of the water table. Consumer choices, which may be guided by a number of factors, especially such socio-economic dynamics as time-savings and distance to the source, and such cultural drivers as habit and taste, have also to be acknowledged. The fact was that, at least until the 1990s, they were, and still sometimes are, ignored.

As already observed in relation to drilling, the Indian rural water supply programme from its inception presented a classic example of a supply-driven approach. The notion of 'supply-driven' is normally contrasted with 'demand-driven'—an economic axis, and one that suggests that programmes which respond to 'demand', or community preference, will be cheaper and more efficient than programmes invented by planners and imposed from on high. During the first decades of the development era, supply-driven water programmes were the norm. Questions relating to economic or environmental sustainability, together with consumer choice and whether there was a match with the socio-economic reality of those they were intended to serve, did not come into the picture. And the reason that the programme was 'supply-driven' was ideological.

During the 1970s, India was still overwhelmingly in thrall to the socialist, state-directed policy idea that the fruits of economic progress would be re-routed to the disadvantaged by means of initiatives subsidized by the state. This was also still the prevailing wisdom of the welfare state in Europe and in many other parts of the world, where those who were disadvantaged—by unemployment, old age, disability, infirmity, or similar problems—were sustained by the re-distributive use of public resources. In India, successive Five-Year Plans tried to tackle the problem of mass poverty by a similar ap-proach on a very grand scale. This helped lead to the expansion of the state, the inflation of the bureaucracy, and a plethora of programmes intended to compensate people suffering from various forms of disadvantage, including food deficit, water scarcity, and drought. Many were beneficial. But the resources available were not always as well or wisely spent as they might have been.

One of the main problems was that there was an underlying assumption that all those grouped under the 'disadvantaged' label were without choices, without skills, without resources, without coping mechanisms, and without viable economies of their own. This was simplistic. It was also damaging. Because it meant that their innate resourcefulness for solving their problems by deploying local solu-tions and resources was dampened or destroyed. They had been assured that 'government would provide' and, cajoled by politicians and bureaucrats, they believed it. To an extent, the government did provide—in some instances to a very remarkable extent. By 1999, for example, three million handpump–boreholes had been installed. According to the coverage statistics, 558 million people, or 75 per cent of the rural population had been provided with publicly-funded water schemes. But provision was one thing. Whether they were used, maintained, and were functioning properly was another.

From early on in the development of the India Mark II, burned by the experience of the past, Unicef focused on the question of main-tenance. When the India Mark II was on the drawing-board, main-tenance and repair were primarily addressed in terms of the necessity for spare parts to be uniform and cheap. The key problem the new pump was designed to solve was the cast-iron pump's lack of stur-diness and inability to withstand wear and tear. While maintenance considerations were incorporated into the handpump design, the assumption was that necessary repairs would be carried out by professional mechanics. What was wanted was a handpump whose durability ensured that mechanical attention during its lifetime would be kept to a minimum. Even if buffaloes were tethered to it, and children bounced on the handle, it was supposed to respond with tolerance. However, it was also appreciated that all pumps break down at some stage and need replacements and there has to be a system for dealing with this.

The earlier experience with the cast-iron pump, when the responsibility had been entrusted to the panchayats and local mistris with disastrous results, surely implied that repairs had to be entrusted to professional engineers. This was the thrust of the institutional maintenance system envisaged for the Mark II. A network of public health engineers trained to mend the pump was needed, as also some convenient way for villages with broken pumps to inform the right officials that they should come to the rescue.

An approach developed experimentally in Tamil Nadu seemed to offer a blueprint with promise. In 1976, the Tamil Nadu Water Supply and Drainage Board (TWAD), with Unicef support, inaugurated a 'three-tier' maintenance system in Tirunelveli and Thanjavur districts. This consisted of a mobile team at the district level responsible for around 1000 pumps, a mechanic at the block level responsible for 100 pumps, and a 'handpump caretaker' at the village level responsible for one pump only (Black 1990). The three-tier system proved a great success in Tamil Nadu and its efficacy seemed to be well established. In 1979 at the national conference on deep-well handpumps at Madurai, it was, therefore, accepted as the countrywide model for handpump maintenance, and training of district teams and block mechanics began apace. Unicef underwrote most of the costs of this programme bar the salaries: training for mechanics, tools, spare pipes and rods, manuals for India Mark II installation and maintenance, and vehicles for the mobile teams. By 1980, 125 mobile maintenance teams were operating in the field.

The third or lowest tier—the 'village handpump caretaker'—was a new cadre of volunteer workers which Unicef enthusiastically promoted. The caretaker was largely the brainchild of Francis, a block development officer who vigorously undertook caretaker training programmes in Tirunelveli district, Tamil Nadu, with all the colour and appeal of a traditional village fair (Jaitley and Daw 1995). At one point, Francis was operating a network of 6000 caretakers. His efforts appeared to have all the ingredients of 'replicability'. The crux of the approach was to confer on users—male and female—a role in the maintenance of village water supply services. The handpump care-taker was not expected to undertake any technical role, other than occasionally greasing the chain and tightening nuts and bolts. The main function was to act as the link with the block mechanic, report-ing breakdowns. Also, the caretaker was supposed to keep the pump area clean and maintain records of the pump's history: dates of repair, actions undertaken, and by whom. Unicef began to support the systematic training of handpump caretakers, first in Tamil Nadu, then in Andhra Pradesh, Orissa, Karnataka, and other states.

At the time of the pump's installation, officials from the block development office would select caretakers for each pump. The person would have to be literate, live near the pump, and preferably have some social standing. Women—key users of the service—were seen as just as eligible as men, and in some states a special effort was made to recruit them. This was the first time that women were recognized not just as beneficiaries, but as active participants in the programme. In groups of between 50 and 100, the caretakers attended a two-day training camp, at which they were taught the pump's basic workings and given some rudimentary instruction in hygiene. At the end, they received a certificate, a log-sheet, a tin of grease, a spanner, and a supply of pre-addressed postcards with which to summon the block engineer. Once alerted, he could respond, and if the repair was beyond him, bring in the mobile district team.

From its earliest adoption as a national strategy, the 'three-tier maintenance' system ran into controversy. Some state PHEDs were not convinced that the handpump caretaker was either necessary or a good idea. Sometimes the numbers of caretakers selected for train-ing was only one-tenth of the number of handpumps installed. This was because of lack of conviction that villagers had anything to contribute to a technical programme. Their views appeared to be confirmed by the high number of drop-outs: a familiar feature of programmes which depend on community workers who are not paid, nor consistently followed-up and motivated to carry out the intended tasks. Unicef, which for the first time in the programme was engaging in a social rather than a technical process, was over-optimistic about what a two-day training camp for people expected to assume an on-going service role at village level could achieve.

But the results were not all discouraging. In Tamil Nadu, the birthplace of handpump caretakers and the three-tier system, maintenance was under the gram panchayats and fully operational (Samanta et al. 1986). In Chindwara and Raipur districts of Madhya Pradesh, Frede Engelund, a Danida Adviser, was as dedicated as Francis in Tirunelveli and able to make the system work in its entirety (Jaitley and Daw 1995). In some districts of Andhra Pradesh, where a strong emphasis was placed by state authorities on training of maintenance personnel at all levels and panchayati raj officials were strongly committed, the three-tier maintenance structure became fully systematized. Some women caretakers formed their own associations, and with guidance from the block development officer, branched out into income-generating activities (Samanta et al. 1986).

At the Social Work Research Centre in Tilonia in Rajasthan, a different approach to handpump maintenance was pursued. Although this was initially supported by Unicef, it quickly led to much-publicized acrimony. Bunker Roy, the activist head of the centre at Tilonia, was a stalwart believer in the capacity of villagers to manage their own affairs. Accordingly he proposed that local school-leavers be trained as handpump mistris in local technical colleges, and that they then be invited to work in a certain area by the panchayats to whom they would be answerable. The task of handpump maintenance would thereby be brought within the socio-economic life of the village. Roy, who persuaded the government of Rajasthan to adopt this as the statewide system, called it 'one-tier maintenance' because the mistri theoretically eclipsed the need for handpump caretaker, block mechanic, and the district teams.

The approach was certainly radical in placing skills, jobs and responsibilities in the community itself. But it turned out that the supposedly 'self-employed' mistris could not operate independently of the PHED, on whom they were dependent for tools, spares, remuneration, and back-up. The one tier of the system was, therefore, an illusion, and the relationship between mistris, panchayats, and the water department unclear and problematic. The system did not work well and every six months, the water department was obliged to carry out a 'repair campaign'. However, this did not prevent Roy engaging in a public diatribe against Unicef and the 'three-tier' system. A war of words took place on the relative merits of one, two, and three tiers of maintenance, and on the virtues of indigenous as opposed to centrally-designed systems invented or influenced by 'experts' with no experience of the realities of rural life (Roy 1987).

Leaving aside criticisms of international organizations and their role in Indian development—some of which have been justly merited down the years—the contrasting positions reflected a sharp ideological divide. On the one side was a belief in people's capacities and the need to avoid bringing in improvements in the name of 'development' which served the bureaucratic and donor agenda rather than the people themselves. On the other, a conviction that improved water supply and maintenance systems were needed and that technical competence and management supervision was the proper role of the government, without which any system would fail. In fact, there was no need for such a publicized polarity of views. Both sides were essentially after the same thing: a maintenance system to ensure handpumps operated effectively. However, it becomes more understandable when one reflects that in those days, standardization and 'going to scale' with universal prescriptions was regarded as the development strategy. Everyone looked for the policy that would unlock the key to the developmental door. As in the technological dimension, so in the social, and if the 'three-tier maintenance' system obtained official blessing at the centre, it might be prescribed for everywhere. Variety and diversity were discouraged. Eventually, the battle over tiers dissolved as states adopted their own variations, two tiers with the block engineer or mobile team as linchpin usually being preferred to three (Mudgal 1997).

In 1986, Unicef commissioned a new survey on the use and condition of handpumps in six states. The overall finding was encouraging. The proportion of pumps operational at any time was 80 per cent, the opposite of the picture 10 years before. However, it was not at all clear that the effectiveness of the maintenance system was responsible. Delays in reporting pump breakdowns were still common with some remaining out of action for months. The high proportion of functioning handpumps was more closely connected to the sturdiness of the Mark II, whose breakdown rate was minimal in the first few years of life. The highest rate of breakdown was in Tamil Nadu. This appeared to be partly because Tamil Nadu's India

Mark II pumps had been installed much earlier than anyone else's and had, consequently, begun to age.

Unicef remained committed to the idea of the handpump care-taker for many years, even though the evidence showed that only in the few places with dedicated block officials and proper back-up did most of these male and female volunteers remain functional for more than a few months. However, there was increasing recognition that any programme or service has a tendency to fail unless the people who are supposed to benefit participate in a meaningful way. Until such time as a better method than handpump caretaking came along for involving local communities, especially local women, in rural water supplies, Unicef in India and elsewhere was reluctant to aban-don it.

During the Water Decade of the 1980s, a new concept began to enter international thinking relating to village water supplies. This hap-pened courtesy of a visionary team at the World Bank, who were determined to promote low-cost appropriate technological solutions to the needs of communities who could not be served by pipelines and sewers. This concept was village-level operation and mainte-nance (VLOM) (Arlosoroff et al. 1987).

VLOM became the big new idea in community water supplies. To begin with, the concept was primarily technological. Handpump development was driven by it for several years. VLOM recognized that it was unrealistic to expect in most rural settings in underdevel-oped countries that an engineering department could be summoned at the drop of a spanner every time a handpump broke down. So pumps should be designed in such a way that, with a little training, any community—especially its women members whose interests in keeping a pump functioning were stronger than men's—could dis-mantle the pump, change washers, or carry out similar minor repairs. In this scenario, the fact that the pump was sturdy was important, but even more important was the ease with which it could be taken apart and mended.

The India Mark II was not a VLOM handpump. In all deep-well handpumps, the cylinder that houses the piston that lifts the water up through the rising main is situated below the water table, far down inside the borehole. The piston seal, usually known as the cup washer, is the most vulnerable part of any pump. In the original Mark II, it was made of leather which was liable to swell with water and jam in the cylinder. Leather is not a uniform material, nor is tanning a uniform process, so that the risks of water-submerged wear and tear on any seal were difficult to gauge in advance. They often perished or played up. In order to change the seal, the piston had to be removed. This could not be done without first extracting the rising main—lengths of galvanized steel pipe—with special lifting tools. Rising main, cylinder, piston, connecting rods—the whole below-ground works—had to be removed, which was quite a skilled operation. In the pre-VLOM days of the Mark II's development, this did not appear to be a disadvantage. No one, with the possible exception of Bunker Roy of Tilonia, expected the people in the community to undertake this task.

However, in response to the pressure for VLOM, in 1982 efforts began to develop a new version of the India Mark II which could be repaired without expensive tools and lifting equipment. As part of World Bank/UNDP efforts to improve them, different handpumps were field-tested in 17 countries, including India. Tests on the India Mark II were carried out in Coimbatore, Tamil Nadu, from 1983–7. One purpose was to test potential improvements to the standard version of the pump to reduce maintenance costs. Another was to experiment with various components to identify and evolve improvements to a basically sound Mark II design. There were questions, too, relating to the use of the pump in lower water table depths, and how to effect the best balance between the diameter of the cylinder, the length of the stroke and the volume of water produced—an important consideration for users. Data were also collected about a number of cost factors, and about downtime, that is, the period when a pump was not in use because of breakdown. This turned out to average 37 days. The findings indicated that, if the pump was able to be mended on the spot, this high level of downtime could be reduced.

The VLOM version of the Mark II—the Mark III—was already in an advanced stage of development. It too was tested in Coimbatore. The key change in the design was to enable the piston from deep in the well to be extracted without removing the rising main. To do this, the rising main had to be wider in circumference than the piston. It would then be possible to lift the piston out of the borehole by pulling up and uncoupling the rods that connected it to the pumphead. Since the use of wider pipes for the rising main was bound to add to the costs of the pump, it was not regarded as a total replacement for the Mark II—at least, not immediately. It would be offered as an alternative, the assumption being that in time it would naturally take over. So a number of minor modifications were proposed for the existing design, including the use of polyvinyl chloride (PVC) or plastic for the rising main pipes (Mudgal 1997). A modification to the material used for the washers and seals turned out to be very significant. This originally came about because the *Sankaracharya*, chief priest of one of the five highest seats of Hinduism, at Kanchipuram near Chennai, wanted an India Mark II handpump for his personal use. A devotee of the Sankaracharya approached Richardson & Cruddas for a pump, but explained that there were religious objections to the use of leather. This unusual demand was a challenge. Earlier efforts to replace the leather cup washer with polyurethane, a synthetic rubber, in the late 1970s had failed. Now an alternative was tried—nitrile rubber. This turned out to be much more durable than leather, and, therefore, came to be standardized for all India Mark II and Mark III pumps.

The modifications to the India Mark II involved an extra cost of Rs 250 (US$ 10) approximately (1991 prices) and the annual savings from breakdowns were expected to be at least Rs 150. The costs of installing the VLOM India Mark III were Rs 1320 (US$ 55) more than the Mark II because of the greater cost of the wider rising main. The maintenance costs would obviously be significantly lower. But it would not be possible to fully recover these costs to compensate for the capital costs of installation for the next seven to eight years (Mudgal 1997). This turned out to be a major disincentive for PHEDs to adopt the Mark III, as did the fact that in very low water table boreholes (100 metres) the rising main became too heavy for the pumphead to support. A smaller diameter rising main overcame this problem, but a smaller piston meant less water was produced by each pump-stroke (10 litres as compared to 12). So users were less pleased with it. As a result, the incentives to switch to the VLOM version of the pump were not very telling. Meanwhile, the improved nitrile rubber cup washer meant that the India Mark II became yet more reliable and enjoyed fewer breakdowns.

So the VLOM India Mark III pump—although a sound and reliable pump costing US$ 25 a year less to maintain—did not turn out to be a universally effective competitor, either for purchasers (states) or for users. In the end, consumer preference is decisive, and VLOM as a technological starting-point for handpump design turned out to be no more perfect a fix for maintenance systems than did the original criterion of sturdiness. In fact, of the two, resistance to breakdown turned out to be more important. As in the case of launching any product into a market, the reasons why the Marks II and III might or might not be acceptable to those who had to make decisions about buying them and using them were not entirely predictable.

During the period of the handpump field testing programme in Coimbatore, the central government was giving increasing attention to drinking water needs. The Seventh Five-Year Plan (1985–90) aimed to provide the entire rural population with 40 litres of potable water per person per day, with an additional 30 litres per head for livestock—cattle, camels, sheep, goats—in desert areas. In 1986, came the establishment of the National Drinking Water Technology Mission. Not only was this to strengthen the programme, but it was also mandated to tackle outstanding issues, including guinea worm eradication (see Chapter 5), water quality problems (fluoride, iron, salinity), and groundwater management (see Chapter 7). A National Water Policy was announced in 1987, giving priority to water use for drinking and emphasizing conservation of a vital resource.

The field-testing of the India Mark II and Mark III handpumps was also expanded so as to gain information about experience with the pumps from other parts of the country. Four large-scale demonstration projects were set up, in the districts of Ranchi in Bihar, Rangareddy in Andhra Pradesh, Betul in Madhya Pradesh, and in four districts in Maharashtra, with technical and hardware support from Unicef and UNDP/World Bank. Although the main emphasis was hardware per-formance, there was also one 'software' objective. This was to develop a village-based maintenance system—the people side of VLOM.

Reports from the field were encouraging. Villagers in Betul district showed an interest in operating and maintaining the India Mark III because repairs were simple and manageable. From Maharashtra came news that the Mark III was repaired with a new alacrity. There was the beginning of a sense that people were taking charge of 'their' handpump instead of regarding it as something belonging to the authorities over which they had no say or control. The first efforts to train women mechanics in maintenance procedures for handpumps began at this time and showed promising results. One setting in which this was tried was in two districts of Rajasthan where, by 1987, an intensive project was underway to eradicate guinea worm. In 1989, 24 women were trained in handpump repairs, and the results were successful (Mehta et al. 1993). But these first steps in community maintenance were no more than that. The policy rhetoric on community involvement ran ahead of practice and of actual experiential knowledge in how to make it work (see Chapter 6).

By the 1990s, another dynamic was pushing the government further in the direction of village-based water supplies management—costs. As the numbers of installations of handpumps and other types of water supply schemes climbed into the millions, it was becoming clear that the state could not afford to bear all the costs of maintenance and repair. At the same time, consciousness in India and internationally was growing that water services were far better used and maintained when local people participated fully in them. They needed to be involved from the start, in planning, siting, and installation, as well as ongoing repairs. In order to gauge their real interest in receiving a new water supply, the simplest method was by sounding out their 'willingness to pay'. Thus, the issues of consumer demand and cost recovery for water supplies maintenance were conceptually lumped together. From both an economic and an ideological standpoint, policy was gradually shifting towards some concept of community management of services.

In due course, these ideas were to take solid shape in the 73rd Constitutional Amendment of 1992, which devolved responsibility for drinking water and sanitation to PRIs. The maintenance wheel was to turn full circle, back to where it had begun in 1971. Would it work better, with the knowledge and maturity of two decades of practice?

❖

The development of the India Mark II, and the extraordinary rate of its take-up all over the country, dominated much of Unicef's effort in water and sanitation throughout the 1980s. Although it sounds a relatively smooth story, and was certainly a 'success story' in its day, the technological ride was often bumpy. No one had any experience of deep-well handpumps for large communities when the new initiatives were taken and key aspects of the design had to draw on ideas that had not been current in engineering thinking for over 100 years. The advantages of modern materials and industrial processes had to be incorporated, and the engagement with local manufacturing on a significant scale as a developmental input was a first for Unicef—in India and the world.

And all the time, the drilling of boreholes was continuing apace, huge numbers of installations were involved, and—never mind the carefully controlled demonstration projects and field-testing—there was constant pressure to come up with refinements and improvements on the basis of educated guesswork, virtually on the wing. Some of these might be tiny adjustments and tinkerings with the size of a nut, the length of a bolt, the grip of a clamp, the turn of a screw. But when millions of pump parts or tools for handling them would subsequently be produced on the basis of one of these design decisions, a poor decision could prove extremely costly, either financially or even in jeopardizing the programme.

This was a classic development conflict between enthusiastic doers and those who prefer careful, scientifically-based, but inevitably slower, forward progress: the old story of the contest between the tortoise and the hare. The fact that, today, the Mark II and III are threatened by a new potential crisis around quality—the original issue of sturdiness—illustrates how yesterday's success may yet become today's incipient disaster. Ultimately, consumer knowledge and satisfaction—with rejection by gram panchayats of shoddy pumps, and retreat to the courts to get satisfaction from incompetent or corrupt manufacturers—is needed to make a meaningful transfer of technology to the community. In most of the districts and blocks covered by the rural water supply and sanitation programme, that situation is still some years away.

By the 1990s, Unicef had also begun to look at other cheap and relatively simple technologies for water supplies than the India

Marks II and III. A number of water-short or water 'problem' environments were not in the hard rock areas. Or if they were, local hydrogeological conditions did not permit community service by handpump–boreholes. Capped springs or other solutions had to be tried. In parts of Orissa, Bihar, and in most of West Bengal, the water table was high or fairly high. At very high water table levels, suction pumps were the rule. But where the water table was below the suction level (7 metres), the solution in the early years of the Indian rural water supply programme was to install an India Mark II handpump over a shallow tubewell. This was a case of expensive technological overkill. The Mark II was designed as a deep-well pump and its mechanism was intended to bring up water from considerably lower depths than the 8, 12, or 15 metres typical of these environments. It operated on the principle of leverage. In addition, the Mark II was not a VLOM pump. Not only were installation costs exorbitant, so too were maintenance and repair.

During the 1980s, a handpump specific for these water table depths was developed in Bangladesh. The first 'tara' pumps (*tara* means star in both Bengali and Hindi) were tested in 1983. Unicef in Bangladesh was closely involved with the Tara pump's development. This pump was a direct action pump and no leverage was involved. By using hollow PVC for the rods connecting the handle to the piston, they would float up when a catch over the handle was released, and very relatively little human force was needed to push them down again. The pump was simple to manufacture, relatively cheap (US$ 150), used components which would be locally available, and was quintessentially VLOM. Its bearing and wearing surfaces were kept to a minimum, and it could easily be dismantled without recourse to lifting tools or sophisticated instruction. It was also compatible with indigenous non-mechanized methods of tubewell drilling and there was no need for casing pipes or other construction complications. It, therefore, cost very little—around US$ 12—to install.

The Tara pump was quickly absorbed into the water supply programme in India for suitable hydrogeological conditions, notably in coastal regions, in parts of Uttar Pradesh, and in West Bengal whose terrain and settlement patterns were identical to those immediately across the border in Bangladesh. After Unicef field-testing, some minor modifications in materials and improvements were introduced.

Once again, standardization and quality control were the emphasis for its Indian manufacturers. Experiences with the Tara, whose installation costs and management were within the capacity of village economies in a way that deep-well handpump–boreholes were not, would turn out to be helpful in promoting community management of services.

Meanwhile, for areas which were arid, where handpumps ran dry in the summer months, where groundwater was chemically contaminated or tasted unpleasant, or where villages were perched on hillsides, a new, or rather ancient, technology began to be advocated—rainwater harvesting. However, like capped springs and gravity flow schemes in hill districts such as Manipur and Mizoram, rainwater harvesting projects were not developed as a mass approach in the way that handpumps were. In most contexts, they were promoted simply because handpump–boreholes were impractical for drinking water and an alternative was needed.

In Nagwan block of Sonebhadra district of Uttar Pradesh, for example, the terrain was rocky and habitations scattered or cut off from the mainland during the monsoon. Handpumps persistently dried up. So, in 1997, rooftop rainwater harvesting was promoted as an alternative. Awareness and motivational activity was an integral part of the project. Workshops involving many types of personnel—health and education departments, anganwadi workers, NGOs—were held. In two phases, 12 villages were selected for implementation. Altogether, 300 ferro-cement tanks on rooftops with a capacity of 5000 litres of water were constructed. Beneficiaries assisted, by helping select the households where tanks were to be constructed and providing the mason with support workers. Costs were Rs 9700 per tank. Some of the masons were women. Within three years, 17 villages in Lalitpur district were similarly covered (UNICEF 2001).

Unicef's interest in rainwater harvesting was, therefore, initially provoked by the need for alternative means to the handpump–borehole or shallow tubewell for household drinking water supplies. But there was another, equally compelling, reason for its promotion—the declining water table and the need for aquifer recharge. In the mid-1990s, that pressing dynamic had yet to gain all the force that would be induced by successive years of erratic rainfall and, in some parts of the country, serious drought. Rainwater harvesting for drought mitigation and, even more importantly, drought-proofing, was yet to make a major entrance on the stage.

REFERENCES

Arlosoroff, Saul (1987), *Community Water Supply: The Handpump Option*, World Bank, Washington, D.C..

Black, Maggie (1990), *From Handpumps to Health*, UNICEF, New York.

Daw, Raj Kumar (2002), 'Submission to e-conference on household water security', No. 69, Orissa Rural Water Project Experience, October.

Government of India (1990), *People, Water and Sanitation: What they know, believe and do in rural India*, National Drinking Water Mission, GOI and UNICEF, New Delhi, based on the report *Water, Environment and Sanitation: A Knowledge, Attitudes and Practices Study in Rural India*, 1988–9, conducted by Indian Market Research Bureau for UNICEF.

Jaitley, Ashok and Raj Kumar Daw (1995), *Contribution of Voluntary Organisations in Rural Drinking Water Supply and Sanitation Programmes in India*, paper sponsored by UNICEF, Delhi and presented to the Water Supply and Sanitation Collaborative Council in Barbados.

Mehta, B. C. (1993), *Involvement of Rural Women in Water Management: Scheme of Women Handpump Mechanics: An Evaluation*, Society for Research Development and Action/UNICEF, Jaipur.

Mugdal, Arun Kumar (1997), *India Handpump Revolution: Challenge and Change*, HTN Working Paper WP 01/97, HTN, SKAT, and UNICEF.

Roy, Bunker (1987), 'Progress with self-respect', article in *Hindustan Times*, December.

Samanta, B. B., Dipak Roy, and T. N. Dutta (1986), *Survey on the Performance of India Mark II Deepwell Handpumps*, Operations Research Group/ UNICEF.

UNICEF [2001] (1985a), *Rainwater harvesting*, field note; Child's Environment Programme, No. 16, UNICEF India Country office.

————— (1985b), *Handpumps for health: Water use in village India* (1985), Brochure based on a 1984 survey, UNICEF Water and Environmental Sanitation Secton, New Delhi.

————— (2000), *Learning from Experience: Evaluation of UNICEF's Water and Environmental Sanitation Programme in India, 1966–98*, Evaluation Office, UNICEF, New York, November, Annexe 6, *The Handpump Story*.

4

Not Just Water,
But Sanitation Too

At the mid-point of the Water Decade in 1985, around 1.5 million Indian children under five still died every year from diarrhoeal disease and dehydration, and half the children aged less than 12 months suffered at least one bout of acute gastro-enteric infection (Black 1990). This huge caseload of childhood illness and death, which included the killers cholera and typhoid as well as other less threatening infections, was the primary justification for Unicef's long-term support to water supply provision. It was also the justification for Unicef's successful advocacy to the Indian government that, wherever possible, handpump water—'safe' groundwater as compared to frequently polluted surface water—be exclusively promoted for drinking.

WHO, the leading UN agency for the Water Decade, set out a large menu of water-associated communicable diseases afflicting people in poor environments. These were classified into 'waterborne', such as diarrhoea; 'water-washed', which included common eye and skin conditions such as trachoma and scabies; 'water-related', such as malaria or yellow fever spread by mosquitoes living in swampy places and transporting the parasite from one person's bloodstream to another's; and 'water-based', such as helminths (intestinal worms). Schistosomiasis (bilharzia) and dracunculiasis (guinea worm) also fell into the latter category because the offending parasites lived some part of their life cycle in water and assaulted the human victim by

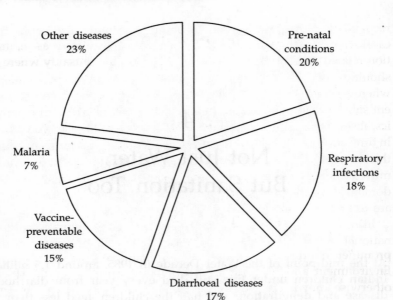

FIGURE 4.1: Major causes of child mortality, globally
[Over 50% associated with malnutrition]
Source: WHO, 1997.

being imbibed in drinking water or entering through the skin (WHO 1992). These water-associated diseases are blamed for 80 per cent of the world's toll of sickness. But this statement promotes a false idea of the problem.

Although water plays a direct or indirect part in their transmission by providing the micro-organisms or parasites which carry the actual illness with a habitat and transportation system, 'unsafe' sanitation practices and lack of environmental hygiene, rather than water itself, are at the root of the spread of these infections. In many cases of sickness, no water is involved at all. Infectious material gets picked up from the landscape or from other parts of the body, and enters the mouth via polluted fingers. In these cases, more frequent washing of hands, even with water that is merely questionably 'safe', would reduce the toll of illness, as does the wearing of shoes outside the home or in any place where germ-laden or parasite-laden material is likely to be present on the ground. Most traditional hygienic behaviours recognize this reality. So it would be more accurate to describe the caseload of disease currently described as 'water-related' as 'sanitation-related', and place more of the emphasis for causality where it should be placed—on reducing contact with disease-carrying agents wherever they are found. Water—that most pervasive and omnipresent substance, moving about under its own special laws and dynamics, absorbing all sorts of substances and particles, not easily contained in time and space—is their favourite element. But it is far from being the only one. All potential environments and pathways for infection must be taken into account if the various types of malady in the domain addressed by 'public health' and its engineering practitioners are to be effectively tackled.

In recent years, there has been more effort by international and national public health authorities to give the role of sanitation the prominence it deserves. At the 2002 International Conference on the Environment and Development at Johannesburg in South Africa, otherwise known as the second Earth Summit, momentum was generated around the target of providing everyone in the world with 'safe sanitation' by 2015. After some resistance from the US delegation, full endorsement for this international target was agreed, alongside the already accepted target of 'safe water for all'. But it has taken a long time for sanitation to gain equivalent importance to water in the international public health agenda. As a result, there is still a long way to go in the catching-up process in terms of official and popular understanding of the key issues, policy development requirements, and sound practitioner experience. To a large extent, this stems from the fact that the public health revolution in the West never needed to de-link water supply from sanitation or human waste disposal because sewerage was integral to the whole public health engineering concept. A household in the industrialized world which has been supplied with water has also been supplied with sanitation, without the less palatable subject ever having to be mentioned. The difficulties surrounding sanitation and its understated—even ignored—role in public health stem, to a considerable extent, from the cultural taboos operating around the subject and the social distaste felt for any mention of excreta. These apply in almost every society on earth, but nowhere are they stronger than in the Indian sub-continent. Thus, in India especially, but also virtually everywhere in the world, it is very difficult to scrub out of the public mind the ingrained idea that providing 'safe' water is the way to deal with sanitation-related diseases, rather than focusing on the more critical interventions of environmental sanitation, including personal hygiene, hand-washing, garbage and solid waste removal, drainage, and, of course, excreta disposal itself.

Every year, the average human body excretes 500 litres of urine and 50 litres of faeces (Esrey et al. 1998). Urine is sterile and harmless to human health. But faeces is full of pathogens and is a highly toxic and health-threatening material. In settings where every household is connected to a sewer, 15,000 litres of safe water per person a year are used to flush away a very small volume of toxic material—a staggeringly wasteful use of water. The water-borne sewerage pipe-lines constructed during the public health revolution in the US and Europe were primarily responsible for rapid improvements in public health. This achievement is much lionized, but is one which the world should now be starting to rue. Sewerage is unaffordable for large parts of the developing world. But that does not stop their municipal administrations, even in relatively poor countries, hankering after sewerage systems and regarding any alternative as distinctly second best. However, progress down the water-profligate sewerage route is necessarily curtailed by its expense. Only 200 of India's 400 major cities are even partially sewered, and only 3 per cent of effluent is treated (Nadkarni 2001).

Instead of confining faecal material to a safe place, most is simply deposited on open ground. From there, wastes are very often flushed into waterways, and whether this occurs naturally by the movement of water in the landscape, or by human intervention in the form of organized disposal or sewer pipelines, the polluting consequences are the same. Indian rivers are awash with raw excreta, and the pollution levels in many mean that they are no longer able to support aquatic life. According to the Planning Commission, 80 per cent of pollution in Indian rivers is from human sewage. Lack of effective sanitation in India is responsible for serious environmental degrada-tion, which could at any time spiral into disaster. In fact, it has already done so. One notorious case was the outbreak of pneumonic plague in Surat, Gujarat, in 1994. The disease had spread from detritus and the rotting carcasses of livestock washed up near the slums after heavy monsoon floods. The epidemic led to a panic evacuation of the city by 500,000 people—one-quarter of its population (*The Economist* 1994). There was a major public health furore, and a quarantine by international airlines and visitors temporarily affected India's economy.

The lack of waste management in India represents a creeping catastrophe in terms of human well-being. The pathogenic material discarded into the open provides a constant health hazard, especially to vulnerable children. Germs and worms passed about on hands and feet find their way via food or dirty utensils to the human mouth, and become the source of most diarrhoeal and other types of infection. Water may be 'safe' when collected from the handpump, but unless its safeness is carefully protected, in the water pot, en route to and in the household, it soon becomes bacteriologically contaminated. If there are inadequate supplies of water for washing bodies, hands, clothes, utensils, and food, 'sanitation-related' diseases easily take hold. In small bodies weakened by poor nutrition, the effect can be devastating. This is the primary cause of death among India's under-fives.

Given its threat to human health, nature has for good scientific reason made faeces an extremely noisome material, which instinctively no human being wants to touch and everyone wishes to have as little to do with as possible. Most cultures, not only in India but elsewhere, are 'faeco-phobic'. Since ancient times, various means have been used by households and communities to dispose of nightsoil. If they feel safe, people have traditionally preferred to go a convenient distance from home to reduce contamination in their immediate surroundings and give fresh air and wind their best chance to perform a deodorizing function. Where possible, the favourite method has been to discard excreta in waterways, using the earth's inbuilt self-cleansing processes to carry it away and neutralize its harmful elements. Another method was to consign it underground, where it could not offend the senses and the micro-organisms could be contained. Yet another method, suited to hot, arid places, was to deposit it in the open air in pre-assigned areas or off the beaten track, and allow the sun to dry it out and kill off the pathogens. In the Rann of Kachchh, for example, special walled areas open to the sky are still established on the outskirts of villages, separately for solid and liquid excreta, and for men and women. The sanitized residues of material deposited out-of-doors can, within limits, be reabsorbed harmlessly into the soil. All these methods were easier to deploy in the country-side than in town, and reasonably safe where population density was low. Population growth, concentrated settlement, and much heavier pollution loads have changed all that.

With the growth of towns and cities way back in historical times, systems of nightsoil collection, transport, and removal to somewhere outside the settlement or the city walls came into being. The work involved—humping and carting foul material—was invariably as-signed to the lowliest members of society. In India, the task was assigned to the lowest of the low. 'Sweepers' were designated a sub-species. Their occupation defined their outcaste status and con-ferred upon them a unique kind of degradation and discrimination—untouchability. Thus in a far distant era, the polluting capacity of faeces was psychologically transferred to the human beings who dealt with them. Their polluting capacity to soil and water was barely regarded at all. One of the first people to draw attention to the effects of this system of waste disposal was Mahatma Gandhi. 'Instead of having graceful hamlets dotting the land, we have dung-heaps. The approach to many villages is not a refreshing experience. Often one would like to shut one's eyes and stuff one's nose; such is the sur-rounding dirt and offending smell.... By our bad habits we spoil our sacred river banks and furnish excellent breeding grounds for flies. A small spade is the means of salvation from a great nuisance.' (Naipaul 1964)

Gandhi was a ferocious advocate for more hygienic methods of sanitation, suggesting on one occasion that 'Sanitation is more impor-tant than Independence'. He campaigned vociferously against the degradation of humanity contained in the concept of untouchability and wanted to end the demeaning life-assigned roles of 'sweeping' and 'scavenging'. The 1947 Constitution, by means of constitutional law and social reform, dedicated the Indian State to the dissolution of the old, oppressive bonds of caste. Gradually, by affirmative action on behalf of 'scheduled castes and tribes', the erstwhile sweepers or untouchables began to abandon their age-old occupation and assume other functions in Indian life.

A minority of the urban population became beneficiaries of sew-erage and septic tanks. But the majority, especially the vast mass of

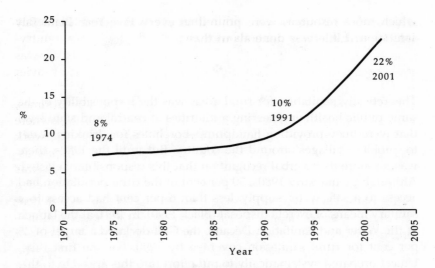

FIGURE 4.2: Percentage of households with toilets

Source: Census of India, National Sample Survey, National Family Health Surveys—
I & II, Multi-indicator Cluster Survey-II.

the rural population, continued in the ways of before. They practised
'open defecation' along river banks, railway lines, streets and paths,
sea shores, and in the open fields. Women, whose modesty required
that they perform this function under cover, sought bushes and
vegetation in the areas around villages and towns, usually at night.
Small children, whose output was seen as less troublesome, squatted
anywhere around the household that they chose, and pigs or dogs
performed some of the sweepers' scavenging and faeces-removal
role.

With the inexorable growth of towns and cities, especially in their
accumulation of squalid peripheries and shanty-towns, and the expansion of the rural population in the post-Independence years, the
need for proper sanitation and cleanliness became more pressing. But
after Gandhi's assassination, in the absence of the most ardent and
celebrated campaigner against pollution as a social and environmental phenomenon, the impetus to do something serious about it via the
formal institutions of government faded fast. Provision of water
supplies in rural areas became a major government concern into

which more resources were poured in every Five-Year Plan. But sanitation? Little was done about that.

Theoretically, sanitation for rural areas was the responsibility of the same public health engineering authorities at central and state level that were busy providing handpump–boreholes for drinking water to 'problem villages' around the country. But until the 1980s, there was no more than verbal recognition that this responsibility existed. Although by the early 1980s, 50 per cent of the rural population had access to a safe water supply, less than 6 per cent had access to a sanitary means of excreta disposal (Black 1990). In 1981, at the launch of the Water and Sanitation Decade, the GOI declared a target of 25 per cent for rural sanitation coverage by 1990. For the first time, Unicef prepared systematically to put effort into this area. Up to this point, its limited involvement with sanitation was some modest support to NGOs for pit-latrine construction, and occasional attempts to bring the issue before the authorities to suggest something be done on a larger scale. To little avail (Wan 1988).

Some of the inheritors of Gandhi's campaign to end traditional practices concerning human wastes—notably Ishwarbhai Patel and Bindeshwar Pathak—had remained active in the intervening years. They and some others continued to carry the torch for the elimination of scavenging, and for the provision of public toilets in urban areas. Alongside Gandhi's *ashram* in Ahmedabad, a showcase 'pit-latrine garden' was set up by Patel to demonstrate the many approaches used around the world to 'dry' sanitation—disposal of human wastes without the use of piped water for flushing, but which may (or may not) use a small amount of water to flush a bowl and land its contents in a pit below the ground. Although these advocates for improved sanitation gained some visibility, partly because they were prepared to be identified with an area of human life that no one else wanted to tackle, they did not engage with the promotion of toilets in rural areas. But they did help to introduce the concept of dry and 'on-site' sanitation technology in the form of 'pit-toilets' to the Indian public. (These were originally known as pit-latrines, a term now being abandoned as derogatory. 'Latrine' implied an unpleasant place, and the term also reinforces the idea of the technology's inferiority to the water closet.) The advantages of pit-toilets were that they were inexpensive, self-contained, did not require a piped water supply for flushing, and they confined excreta hygienically. Their disadvantages were that they were culturally disliked as claustrophobic, that people wanted excreta to be deposited at a distance from their homes and not kept under the ground beneath their feet, and that, when the pit was full, it had to be emptied and the content handled—by someone.

Early in the Water Decade, on an ad hoc basis, Unicef became more involved with various NGO community-based sanitation programmes, mostly in poor urban areas which, at this stage, were the exclusive focus of sanitation efforts. The hope was that the organization could repeat the role it had played in the evolution of the water supply programme and act as a bridge between pioneering, often small and isolated NGO initiatives, and the adoption of a service delivery approach by central and state governments. One of the early principles to emerge was the need, strongly emphasized by the NGOs, that any enterprise around sanitation required community participation and must be built on demonstrated consumer demand. Their experience showed that the provision of a free toilet without any attempt to motivate recipients virtually guaranteed that it remained unused (Wan 1988).

Another feature of the NGO projects which Unicef took up was to broaden the definition of 'sanitation' to cover all aspects of personal hygiene, waste disposal, and environmental cleanliness which could have an impact on health. Sanitation in the form of garbage removal, clean paths, drainage of stagnant puddles; and personal hygiene in the form of teeth-cleaning, nail-clipping, and washing hands with soap, could be promoted independently of toilet construction. So educational materials were produced on the connections between dirt, water, and disease, covering personal and domestic hygiene, vector control, food cleanliness, drinking water storage, and the use of smokeless *chulhas* (mud stoves). This was Unicef's first foray into a purely 'software' water/sanitation venture unconnected to any kind of technical training programme. The materials were disseminated to NGOs, community development officers, school teachers, anganwadi workers, and other social-programme personnel whose primary targets were women and children.

In 1983, Unicef became involved in a Technical Advisory Group (TAG) led by the World Bank and including members from the government and UNDP. Finally, here was an attempt to develop a sanitation approach for the country and match some of the energy and resources expended on water in a parallel effort for sanitary living. Regrettably, the TAG saw sanitation in purely technical, engineering, and construction terms. Perhaps it was difficult enough to persuade professional engineers from the central government and state ministries that they must take seriously 'on-site' sanitation and the lowly pit-toilet as the only feasible technological option for poorer communities; that there could be no question of universal sewerage, and that such communities could not simply be ignored. There was, too, the example of rural water supply which implied that if a 'model' could be produced—a standardized 'hardware' package, training for its implementation, and the rest—sanitation in the narrow form of toilet construction for the safe disposal of human excreta could 'take off' in the way that handpump–boreholes had done (UNICEF 2000).

Unfortunately, the TAG, despite Unicef's membership and its significant financial input, did not manage to incorporate the hardware lessons about appropriate technologies, maintenance systems, and customer acceptance from the handpump experience or from its support to NGO sanitation projects. The most glaring mistake was the failure to understand the basic parameters of life in typical rural areas. The farming family they had in mind—if they had one in mind at all—was in the upper income landowning category, living in a large permanent dwelling. The majority of people did not live like that. Their houses were simple, often made of natural materials. The concept of a toilet was itself a strange one, and a separate stand-alone 'room' built of brick and mortar would never be accepted for this purpose. Ignoring all of this, the solution the TAG exclusively proposed was the twin-pit, pour-flush toilet.

The ideas dominating the development of this toilet were technological. It should be an advanced and superior version of the humble latrine. One of its key features was the use of two pits, one of which could be closed off while the second pit was used. In this way its contents would be rendered into harmless composted material and no one need handle the obnoxious waste (Wan 1988). The second key idea was the 'pour-flush' feature. Water was to be used, in minimal amounts, for personal cleansing and for flushing. As in any flush toilet, some of the clean water poured down it was held in a U-shaped 'trap' below the pan, acting as a barrier between the user and the contents of the pit. The pan would be of smooth and easily-cleanable material, such as ceramic. Finally, it was thought that this should be a pleasant-to-use toilet, not a small, claustrophobic cubicle. It should be sturdy and respectable, not cheap and inferior. But the problem was that these features made it expensive to build—more than the cost of many people's entire dwellings. The pour-flush, twin-pit toilet cost over Rs 2000 (1980s prices). It was built throughout with brick and mortar, including the superstructure.

This was the sanitation model adopted by the government's Centrally Sponsored Rural Sanitation Programme (CRSP) when it was launched in 1985 and began to get underway in the following years. Unicef, which claimed to be aware of the pitfalls of a one-size-fits-all technical fix, and of the need to incorporate motivational messages, health education, and demonstrable community participation into any sanitation programme, failed to influence in these directions either the recommendations for the strategy or its implementation on the ground. The programme had a poor impact from the start, one from which it has ever since been trying to recover.

No one disagreed that the twin-pit pour-flush toilet was an excellent facility. It was indeed the palace of latrines. That was its problem. It was way beyond the means of the vast majority of India's rural population, even if they had been first motivated to want one, which they had not. The idea of the rural sanitation programme was that the toilets would be provided free to low-income households, particularly those belonging to scheduled castes and tribes, and somehow everyone else would manage without or be inspired to build their own. As with rural water supplies, the policy was consistent with the ideology then driving India's planning process—that the way to mitigate poverty was for the government to provide free or heavily subsidized services. Therefore, no consideration was given to any form of contribution from recipients, nor to the need to promote its virtues for health. Although the NGOs suggested that local masons could be trained to undertake toilet construction, the idea was not taken up. Contractors would be hired by state departments

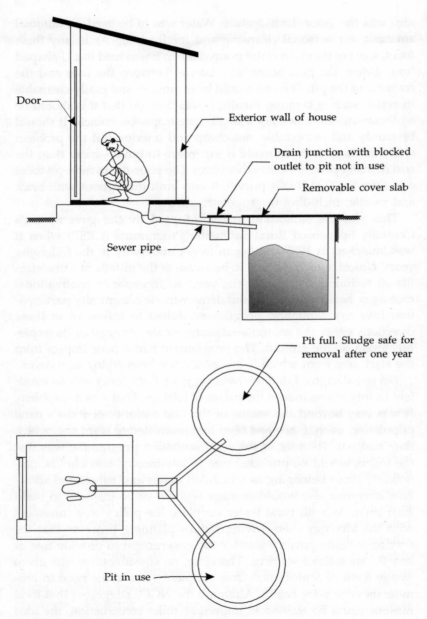

Door

Exterior wall of house

Drain junction with blocked outlet to pit not in use

Removable cover slab

Sewer pipe

Pit full. Sludge safe for removal after one year

Pit in use

FIGURE 4.3 (a): Twin-pit pour-flush toilet

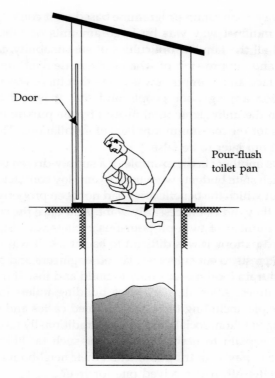

Door —

Pour-flush
toilet pan

FIGURE 4.3(b): Single-pit pour-flush toilet

Source: Adapted from: Water Engineering Development Centre (WEDC) Loughborough, UK.

in the time-honoured way and follow, also in time-honoured ways, the technical blueprint provided. What usually happened was that, since the budget was only enough for one or two latrines to be built in each village, they were constructed in the most influential households. Where they were constructed in the compounds of poorer villagers, they were usually used for storage of grain or other valuables. Few villagers other than well-off landowning families would think of using such a structure for defecation.

The programme was duly taken up by state engineering departments, partly due to aggressive promotion by some TAG members, and incorporated into mainstream public health engineering practice.

The only way a sanitation programme based on it could 'go to scale' even in a minimal way was by large amounts of subsidy, which introduced all the familiar difficulties of sustainability, inflated bureaucracy, and suppression of whatever real demand for sanitation existed. In fact, an assumption was made that there was no demand for sanitation among poor people and that it would have to be imposed on the unhygienic rural masses by the pursuit of construction targets for the maximum numbers of installations. This assumption turned out later to be false.

Here was another classic example of a supply-driven programme. One in which state budgets were used to employ contractors to build things about which the beneficiaries had not been properly consulted and which they did not necessarily want. And once the construction of a target number of twin-pit pour-flush toilets was established as the standard pattern, it was difficult to backtrack. It was not in local officials' interests to cut subsidies, lay off engineers and contractors, and instead train local masons how to build and install more affordable alternatives. After all, they were building toilets for poverty-stricken people, including those in scheduled castes and tribes—the disadvantaged whom social programmes traditionally targeted. How would one explain to new applicants for such facilities that they would have to pay a contribution, when their neighbours, who were probably better-off, had received one for free?

More than 15 years later, some public health engineering bodies are still building edifices to better excreta disposal in their districts with sublime indifference to their cost-effectiveness and use. There may be more commitment to promoting hygiene in home and community as integral to a healthy way of life, but implementation in many states or parts of states is far from adequate. A recent study showed that too high a proportion of government-subsidized toilets are not used for their intended purpose, but instead have been converted into store-rooms or kitchens—household improvements that they value more highly (Saxena 2003). This is an indication of a slippage between the provision of services to a target group, and the establishment or creation of a demand for the service by the intended users.

Although Unicef failed in the early stages of the centrally-sponsored sanitation programme to persuade the government to take a more holistic view than mere toilet construction, or to build on any of the other lessons of the NGO projects it was supporting, it began increasingly to promote these on the ground. As Unicef stepped up its own involvement in sanitation by appointing sanitation officers to its staff in state offices (most assumed their functions in 1987) (Wan 1988), these principles dominated pilot schemes and other 'total sanitation' ventures on which they locally embarked. At a certain stage, Unicef withdrew its support for fully subsidized toilets under the centrally-sponsored programme. Instead, it backed local activities in states which were imaginative about the non-construction, or software, aspects of environmental sanitation. The hope was that, eventually, the supply-driven, toilet-only, one-size-fits-all central sanitation policy would be revised.

The water-rich landscape of West Bengal is luxuriant and crowded. Brilliant green paddy fields are interspersed with clusters of thatched and tiled roofs nestling together under shady palm trees. Every square inch of this fertile and steaming landscape is spoken for, mostly by modest farming families, some—but by no means all—of whom are extremely poor. In the villages, narrow paths skirt compounds and ponds, fruit trees, and plots of vegetables. No house is very far from the next. Some are elaborate, with two storeys, carved and decorative features, and substantial shaded forecourts. Others are tiny. In such houses the family sleeps together in a room barely large enough to hold a bed, cooking out of doors in a 'kitchen' consisting of a few stones, some burning embers, and a steaming pot. This is garden suburbia with many different social groups intermixed, the closest thing in the Indian countryside to cheek-by-jowl, socially accommodating, urban living.

Medinipur district, with 9.6 million people—nearly one for every 100 Indians—is the country's most populous.[1] It is also one of the country's lowest-lying areas, prone to floods during the monsoon,

---

[1] Medinipur was split into two districts on 1 January 2002, Purba Medinipur (East) and Paschim Medinipur (West). The total population figure is the 2001 Census figure; it was 8.3 million in 1991. During the key period of the project described here, Medinipur was one district.

and also to epidemics of cholera, typhoid, and other sanitation-related disease. The spread of such infections is greatly assisted by open defecation, which, as recently as the 1991 Census, was practised by over 95 per cent of the rural population. In the late 1970s, an NGO, long active in the more deprived parts of West Bengal, including Medinipur, the Ramakrishna Mission Lokasiksha Parishad, noticed that the provision of handpump drinking water, nutritional supplements, and immunization had failed to make any impact on the health and survival rate of children under five. Further study revealed that the lack of sanitation was responsible.

India abounds in home-grown organizations with very particular characteristics to which the all-encompassing label 'NGO' does not do justice. The Ramakrishna Mission, based at Narendrapur close to Kolkata is an organization with its own spiritual cadre, as well as an extensive network of dedicated development outreach workers and volunteers. The Mission is devoted to spreading the inspirational words and the deeds of Swami Vivekananda, follower of Sri Ramakrishna, a nineteenth century sage. Swamiji, who was active at the turn of the twentieth century, had a particular theology of development which was quite visionary for its time. He believed that rural people's own skills and capacities should be built up so that they could stand on their feet, pursuing their own salvation through hard work and spiritual strength.

The Swami's devotional successors would not be likely to couch this in political terms, but it nonetheless represents a very different vision of development to that of Jawaharlal Nehru and his political successors. Its ideological connections are much closer to Gandhi's ideas of self-development and enlightened village rule. The objective of both spiritual and developmental personal growth is synonymous with what today would be termed 'empowerment', for the Ramakrishna Mission tried from the outset to change mindsets by a strong emphasis on education of all kinds and to build viable democratic institutions at the grassroots. This was not envisaged as an alternative to government service provision but, on the contrary, as an effort where the two should come together and work in tandem.

The Mission's strategy was to reactivate and restructure local organizations, especially women's and youth clubs, democratizing them, and training their members. They gave spiritual direction to the Swami's moral teachings as well as practical inputs: management skills, non-formal education, vocational and entrepreneurial expertise to discourage young people leaving the moribund rural economy for the bright lights of Kolkata. The organizations were registered as legal societies and bound together in 'clusters', so that there was a representative hierarchy for self-management and supervision. These interacted with the panchayat institutions—strong in West Bengal—and the administrative bodies: block officials and district magistrates.

These networks of organizations, especially the youth clubs, were part of the essential groundwork for the intensive sanitation programme launched by the Ramakrishna Mission in 1990, with Unicef support and state government approval. So was their previous experience with toilets. From the early 1980s, the Ramakrishna Mission was one of the NGOs whose sanitation efforts Unicef had backed. To begin with, Unicef provided technical training for constructing the twin-pit toilet, and a subsidy of 60 per cent (Rs 1200) towards every household toilet to the level of the squatting plate (the superstructure to be built by the customer). In spite of its high cost, the Mission managed to mobilize some genuine local interest in the toilet. This led to a request for 350 units from the Mothers' Committee of Arapanch in South 24 Parganas. Suddenly, Unicef found itself caught short with a lack of funds. This forced a key change of dynamics. The Mothers' Committee of Arapanch agreed that the reduced amount Unicef could offer should be spread more thinly, and subsidy would be given for 40 per cent, not 60 per cent, of the costs so that all 350 toilets could be built.

The Ramakrishna Mission became convinced as a result of this experience that toilet provision could be done with no subsidy at all, so long as educational motivation through community organizations was effective. No organization was better placed to try out this strategy and make it work.

The proposal submitted to Unicef for 'intensive sanitation' in Medinipur broke with every conventional view of what was practicable until this time. The starting point was that demand could be created for toilets. They were not a dead letter in most rural Bengali minds, especially in the minds of women who treasured the privacy and round-the-clock availability a toilet could provide in the crowded rural environment. Although in the first two years of the project relatively few toilets were built, thereafter, attitudes began to shift rapidly. By late 1994, the project had reached more than 2600 villages (UNICEF 1994). Over 52,000 toilets had been built and no subsidies had been provided at all.

This breakthrough was achieved by putting first priority on awareness-building and mobilization, and second priority on technology and construction. The Ramakrishna Mission conducted motivational camps and instruction sessions for all kinds of personnel: door-to-door motivators for each village, village masons, local mistris, singing squads, youth clubs, wall painters, and members of the local panchayats. The effort showed that age-old habits, thought to be intractable, could be dislodged if persuasion was persistently and repeatedly undertaken. The door-to-door animators, who visited a 'beat' of neighbours, found that it took an average of five visits for persuasion to pay off. And pay off it did. People, especially women, were prepared to abandon open defecation and to become evangelical about its ills. The cleanliness of the village—the absence of faeces on local paths, the removal of garbage—and its fitness to receive visitors became a status symbol.

Members of the youth clubs were the original motivator group, each visiting 100–200 families. The clubs provided an interest-free loan for a toilet if the customer put down half the price. Production centres for toilet slabs, pans, and water-seal traps were set up. Masons were trained to manufacture slabs (usually men), and mosaic pans with a white cement base fitted with water-seal traps (women). Toilets were offered at a range of prices. The cheapest, the one that most people selected, consisted of a basic slab, pan, and trap to be used to cover a single unlined pit, and cost around Rs 300 (Rs 400 by 2002). When circumstances allowed, a customer could upgrade to a brick-lined pit (Rs 710), or to twin pits. The twin-pit pour-flush with ceramic pan and a brick toilet cubicle was at the top of the range, costing around Rs 3000 (Rural Development Department, West Bengal 2002).

Gradually, a new local employment, manufacturing, and sales sector developed around a previously unwanted consumer item. At the same time, a 'whole sanitation package' including education about safe handling of drinking water, food hygiene, the methods of disease transmission, solid waste disposal, and the need for environmental cleanliness had been transmitted. By 1995, 4686 out of around 10,000 villages in the district had been reached, with the involvement of over 1000 youth clubs and 11 cluster organizations (UNICEF 1996). From 12 per cent coverage of toilets in 1990, the project set its ambitions on reaching 80 per cent by the end of the decade.

The West Bengal government always took an interest in the project. But the authorities became much more actively involved after the 1992 passage of the 73rd Amendment to the Constitution, devolving many governance responsibilities to the local PRIs and giving women a role in local government. Towards the end of the 1990s, after an intensive literacy effort in Medinipur had begun to near its conclusion, the district authorities, working with and through the panchayat institutions—especially the gram panchayats and lower level *gram samsads* (local councils)—began to throw their weight behind the goal of 'saturation' toilet coverage. Their ringing official endorsement of not just the benefits, but the necessity, of using toilets proved decisive.

Fast forward to 2002, and we discover that the groundwork set in place by the Ramakrishna Mission has become the platform for an extraordinary success—for which the Mission credits the government and the panchayats of recent years. Out of 25 blocks in Purba Medinipur (East), 100 per cent coverage with household toilets has been certified in Nandigram II block, claimed in three more, and is close to achievement in another eight. This is a 'first' in India. Five or six further blocks have toilets in 70–80 per cent of households. The District Magistrate, Anil Verma expects the whole district, with a population of 4.5 million, to be 'saturated' by 2004.

This remarkable achievement has only been possible with household-by-household persuasion. 'The final 10–15 per cent were the most difficult group,' says Anil Verma. 'We had to push very hard with extra explanations and motivations. We said: "If you don't cooperate we won't help you in other ways."' The District Magistrate himself went to visit some of the recalcitrant families, tirelessly doing his rounds until late in the evening. And the gram samsad was not above sending a policeman—not in his uniform—to have a persuasive chat. Of course, having built a toilet, some family members may still be reluctant to use it. But if they are spotted defecating out of doors, they are given a difficult time by their peers. People have realized that even one uncooperative defecator can prejudice others', especially children's, health. It has become a point of pride to become the first district in the country to abandon the bad old insanitary ways—universally and for good.

The special network of community organizations created by the Mission, and its own supporting network of Lokshiksha Parishad field workers, has encouraged the widely held view that the sanitation success in Medinipur is a 'one-off'. Without the painstaking work the Mission undertook in the pioneering years, the mass take-off which followed government endorsement in 1992 would not have materialized so rapidly. As in so many community-based endeavours, the nature of the NGO and its form of intervention in the early years was, indeed, critical. Sceptics seize upon this aspect to say that the model will not be replicable. But although it is not easy to develop a replicable model for what is essentially a social programme, it is not impossible, especially where the programme is to be replicated in an almost identical social, geographical, cultural, and economic environment.

In the early stages, the intensive work of a community-based NGO network may be vital. But when there has been experience in solving many of the problems, and the critical turning point in social behaviour and attitudes has been reached, the hardest row has been hoed and expansion can move ahead more smoothly. There will always be potential traps and pitfalls, but the basic parameters are established. At this juncture, assured commitment from administration officials and political leaders and sufficient budgetary support can spread a programme rapidly. In this sense, a development programme is not unlike any entrepreneurial venture: the most difficult stage is the first. The reason why so many development schemes do not 'take off' is that the early stages are not adequately accomplished and activities have not become grounded in community desires and institutional structures before 'scaling-up' is over-enthusiastically attempted.

In the last few years, with Medinipur's progress assured, Unicef has been actively promoting the sanitation programme in other districts of West Bengal. The strategy for 'scaling up' the programme to the whole state, consisting of 80 million people, is to identify other NGOs which can take the lead in the communities, working with the PRIs and the authorities in the way pioneered in Medinipur. In

Howrah district, an NGO called Ananda Niketan which is well established and has many experienced staff, has taken the lead in certain blocks, with 10 other smaller NGOs assuming responsibility in others. The NGO personnel, after attending courses funded by Unicef at the State Institute of Panchayats and Rural Development near Kolkata, train local masons and animators in the villages.

Gradually, leading NGOs in all West Bengal districts are embarking on the business of building a demand for toilets, providing the product, and embedding sanitary ideas into the local economy and attitudes. The pattern for setting up production centres under their auspices is uniform; the range of products and the price structures; the incentives and stipends for motivators; the motivational and educational materials. In Howrah, the district authorities are very committed and have allocated considerable funds to the local panchayats for the programme. So the rate of toilet construction has been rapid. Howrah is already on target to reach 'saturation' coverage not much later than Medinipur. The aim for the state as a whole is to manage full coverage with household toilets by 2008. Unicef, for one, thinks it can be done.

Nothing succeeds like success. But as yet, the household sanitation success in West Bengal has not proved attractive enough for other states to copy even those, such as Orissa, with similar terrain and a comparable socio-economic profile. In Orissa, rural sanitation coverage is only 8 per cent. One important difference is that the PRIs of the state are a pale shadow of the dynamic ones of West Bengal. Even in states such as Karnataka and Andhra Pradesh, where sanitation in schools is becoming popular, rural coverage with household toilets is still only 17 per cent. In some states, it is much lower: less than 9 per cent in rural Bihar, 6 per cent in Gujarat, 11 per cent in Rajasthan (UNICEF 2000). Where it is higher, over half the facilities are not used (Saxena 2003). So 'going to scale' with household sanitation in West Bengal does not seem to be easily replicable.

There is another note of caution. 'Saturation' coverage with toilets, many of them with unlined pits, and in an area vulnerable to flooding, has never been done before on such an intensive basis. There are already probably close to a million pit toilets in the West Bengal countryside, and there will in due course be hundreds of thousands more, all containing faeces in some condition of risk. Tests have been carried out on the soil, to see at what distance from a pit pathogens can leach through it. Theoretically, no pit is dug within that distance (five metres if the soil contains more than 25 per cent of clay which is normal in Medinipur and in most of West Bengal; 10 metres otherwise) from a pond or other water source used for drinking, such as a dugwell. But in the enthusiasm to achieve 100 per cent household toilets, will such important parameters for health be 100 per cent observed? Nobody knows what the outcome potentially is of confining so much raw excreta to the ground in such a waterlogged environment.

There is always the chance that the very success of an approach will throw up problems which had been difficult to anticipate in the early stages—as Unicef has frequently experienced. All of those—the government, NGOs, PRIs—who have been involved in the sanitation success of Medinipur will be carefully watching to ensure that it does not become tomorrow's setback. A different approach to sanitation in high water table areas is now in the early stages of being explored—ecological sanitation. A project in coastal Kerala has experimented with toilets designed to permit the separation at source of liquid from solid excreta. The faeces are kept separate and safely enclosed in a small brick chamber. After they have dried out for some months, they become an entirely unobjectionable substance that can be withdrawn and used as a soil compost. Meanwhile, the sterile and nitrogen-rich urine can be led away from its container by a pipe and used for fertilizing a garden. The advantages of 'dry sanitation' are immense from a health and environmental perspective. Water is not squandered for waste transport. Faeces are confined and not left in the open to pollute ground and water; public health risk is minimized; at the same time, the natural fertilizing qualities of human wastes can be deployed for gardens and orchards.

The problem to be overcome with ecological or dry sanitation is cultural acceptance. Many sceptics believe that in India, the approach will confront enormous resistance. However, the taboos which once faced any sanitation programme in India have gradually crumbled in the face of growing public awareness of health and hygiene. Unicef is now backing some pilot 'eco-san' projects in Bihar and some other high water table or flood-prone locations. The dry sanitation approach may still, at present, be hoeing a difficult row. But in environments such as West Bengal, where women and community leaders are regularly confronted by monsoon-flooded toilet pits and a filth-filled landscape, there is a strong possibility that, in time, ecological sanitation will find favour.

By the early 1990s, the idea of using 'motivators' and social mobilization techniques, so central to the Medinipur sanitation strategy, was no longer novel. It had been used to good effect in disease-control campaigns, especially the eradication of guinea worm (see Chapter 5). However, the idea of village production centres employing local artisans to manufacture toilet pans and slabs—also called Sanitary Marts—certainly was. This idea was exported to other states in an attempt to replicate the Medinipur experience.

In West Bengal, the economics of the Sanitary Mart are based on the premise that the toilet has to be fully affordable for everyone except those 'below poverty line' who will be entitled to a subsidy of Rs 200 from the government towards its installation, and that the unit of choice will, therefore, be the simplest version of slab, pan, and trap. These are cast in moulds using cement. In the pan, white cement and quartz chips are added so that, when ground down, a smooth whitish surface improves its look and the ease with which it is cleaned. Thus the toilet components are relatively easy to make and well within the capacity of a local mason, male or female. The toilet is serviceable but far from grand. Within the selling price of each unit, Rs 40 is split equally between the Mart and the motivator responsible for that customer's purchase. Thus profits and incentives are built into the centre's operation, and there is a careful fit between the economic reality of customer demand and the economic reality of the technical intervention.

The role of Unicef is to provide a start-up fund to allow the implementing NGO to get the centre running: hire premises, purchase moulds, employ part-time managers, take in a stock of materials, and set up the production centre. Unicef also provides training for NGO leaders, Mart managers, motivators, and masons. NGOs are also provided with information, education and communication (IEC) materials and motivation kits. Once on its feet, the centre or Mart should be self-supporting. However, not all can be so. As in any chain of outlets producing and selling similar goods and operating along similar lines, some do not succeed. But more than 75 per cent are self-supporting and around 40 per cent are doing very well indeed.

When Sanitary Marts were first set up in West Bengal, they were conceived as shops as well as production centres. Along with toilet components, soap, tooth brushes, brooms, nail cutters, cheap footwear, domestic water filters, alum, bleaching powder, oral rehydration salt (ORS) packets, and smokeless chulhas were also stocked. These items all belonged to the concept of the total sanitation package. In 1993, the Drinking Water Mission adopted the idea of Rural Sanitary Marts as a way to boost the central sanitation programme (Dixit 1998). By this stage, the policy-makers had accepted that sanitation should be seen as a package of environmental and household cleanliness measures, not simply as toilets; and that the promotion of sanitary behaviour was an important adjunct to the subsidized government scheme. The Rural Sanitary Mart was seen as one way to promote sanitary items and, thereby, promote sanitary behaviour, and also as a sales outlet for toilets. The assumption on which it was based was that some demand did exist for sanitary toilets, and there was a gap in the market which was failing to cater for this demand. The promotional literature put out by the Rajiv Gandhi Drinking Water Mission read: 'Establishment of a Rural Sanitary Mart is a step towards commercializing the provision of sanitary facilities (UNICEF and GOI 1991).' Unfortunately, the step turned out to be rather small.

The Rural Sanitary Mart concept was tested in the field by a technical institute, the Institute of Education and Rural Technology in Allahabad district of Uttar Pradesh. The first Sanitary Mart in Allahabad was set up in 1991. Within two years 16 were operational (UNICEF 1993). Unicef provided technical support and the funds to get the Marts going. They were supposed to be a one-stop retail outlet dealing with all the materials needed for the installation of a toilet, an advice centre for interested households on all matters to do with environmental cleanliness, and they carried a list of recommended masons able to undertake construction of the different models of toilet shown at the centre (Samanta 1997). The Marts were expected to become self-sustaining within two years, tapping into latent demand for low-cost toilets and helping to create more sanitation interest.

There was great enthusiasm from Unicef and the government for the Sanitary Mart as the hub of a new no-subsidy household sanitation approach. Even before the Allahabad demonstration programme had proved the Marts' commercial viability, they were set up in many other places. Too many, too fast, too optimistic: the three-fold blight of so many development interventions. Little time was spent analysing difficulties and key parameters for a successful retail and service venture in advance of 'scaling-up'. And, unfortunately, the commercial prospects for toilets turned out to be less than had been assumed. Most Marts might sell an occasional toilet seat, but the incentives and profits were not enough to build a retailing business on sanitation, let alone one which acted as an information and advisory centre.

Many found that the customers that did present themselves were more interested in ceramic pans and permanent installations. Without the intensive motivational effort and one-on-one persuasion tactics employed in West Bengal, there was no ready queue of lower-income customers lining up to order. The Marts were often poorly sited—not in main shopping markets—and often had only a single manager to staff them. Many that tried to make retailing viable by selling hair brushes, cleaning utensils, toothpaste, soap, shaving cream, and other wares, found themselves competing unsuccessfully with local shops who already stocked these items. Managers reported many problems, and those with business experience commented that it was much easier to make a living from running a tea stall or grocery shop (UNICEF 2001). Only where local development officials put real effort and funds into the programme, training motivators, masons and panchayat officers, setting up Water and Sanitation Committees, and running intensive mobilization campaigns, did the Sanitary Mart go far. There were cases where determined entrepreneurs made a success of a local Mart. One such case was in the small market town of Nabhi in south-east Uttar Pradesh. Here, with training from a local NGO, an interest-free loan of Rs 50,000, and further investment of Rs 200,000 from his brothers, 31-year-old Sri Yadev made his Mart into a thriving business with a turnover of over Rs 300,000 a year (UNICEF 2001). But 70 per cent of this turnover came from the sale of paint, hardware, and electrical items, and 20 per cent from soap, shampoo, and buckets. If he sold 100 toilet seats in a year, he considered he had done well. This was clearly not evidence of a breakthrough in the popularization of sanitation in rural areas, nor did it show that there was more than a very modest existing market—in Sri Yadev's view, one dependent entirely on women's desire for privacy and their families' economic status—to be tapped. In spite of promotional efforts, it would take time for any significant change in attitudes.

Belief in what had seemed such a promising idea was difficult to dislodge. Only 100 Sanitary Marts had been set up by 1994, with a total countrywide output of 17,000 toilets. Despite this discouraging evidence, Unicef insisted in 1995 that the Sanitary Mart be adopted as an integral part of any sanitation project in which it was involved. Progress continued, but slowly. By 1998, 450 Marts had been set up throughout the country. But many soon ceased to operate in any meaningful fashion, or simply went out of business. In West Bengal, the parlance of Sanitary Mart was dropped in favour of Village Production Centres, and the retail side for household hygiene wares virtually disappeared. The difficulties of spreading demand for sanitation could not be overcome by setting up semi-subsidized toilet-promotion stores. With reluctance, it had to be admitted after a decade of trying that this idea as the centrepiece of a countrywide rural sanitation strategy had proved wildly over-ambitious.

However, in other areas there had been significant advance. In 1992 during preparations for the Eighth Five-Year Plan, a National Seminar on Rural Sanitation was held. In the previous five years, Unicef had been pursuing its small-scale efforts in sanitation through NGOs such as the Ramakrishna Mission in Medinipur, and others in Rajasthan, Tamil Nadu, Haryana, and Karnataka. These had all promoted a package of 'total sanitation', and communications and social mobilization were at the centre of their strategy. Women's groups often provided the critical channel. These projects were vital in enabling Unicef to fund, test, and advocate a variety of approaches. Its field offices, NGOs, and state partners explored existing attitudes and practices, tried out methods of cost recovery and community involvement, and monitored closely how interventions worked—or didn't. An organizational approach to sanitation was developed, based on solid research and experimentation. When the 1992 National Seminar took place, Unicef and state representatives who had seen the results of the local programmes were able to use the lessons learned to advocate a major change of sanitation policy. Years of groundwork and trying to prompt a change of heart were finally paying off. The renewal of the earlier policy so dearly needed was fully set in motion.

In 1993, the central sanitation programme issued a new set of guidelines reflecting the changes Unicef, NGOs, and state partners had proposed. These were based on the Medinipur experience, and the subsidy for home toilets was reduced from Rs 2000 to Rs 500. The idea of the rounded package—toilets, wastewater disposal, food hygiene, home protection of drinking water, clean paths, personal cleanliness, and garbage disposal—was endorsed. There was a shift from 'hardware' to 'software': not so much construction of brick-and-mortar amenities but more emphasis on information, education, and communication to promote health awareness and build demand. Instead of one 'this is it' toilet design, there was a range of options more suited to the modest consumer purse; and instead of a single 'this is it' programme delivery structure, there were now several options: sanitary marts, revolving funds, NGOs, the private sector.

Thus, the direction of the programme was gradually changing towards 'let a thousand flowers bloom', abandoning the old-fashioned notion of a solution imposed from on high. In its forward-looking thinking, the ideas it reflected were very much those which had become current in international discussion on water supply and sanitation during the early 1990s: the insistence on community participation in schemes, the emphasis on women's roles, the involvement of the private sector, the importance of software. These were all themes circulating in the international firmament, crystallized in the International Conference on Water and the Environment at Dublin, and in the first Earth Summit at Rio de Janeiro, both in 1992.

However, there was another important dynamic at play—India's own changing policy climate. The Eighth Five-Year Plan (1992–7) was to devote Rs 6740 million to sanitation, 11 times more than its predecessor. During discussions on how it should be spent, the reality had dawned that even this amount would at best cover 10 per cent of households below the poverty level unless subsidies were slashed and policies altered. In 1999, the central sanitation programme was recast as the 'Total Sanitation Campaign' (TSC) with the idea that advocating the take-up of sanitation was much more important than building actual facilities for people. Instead of full or 80 per cent subsidy for scheduled castes and tribes, there was now no subsidy for those above the poverty line, and even those below it were to pay 20 per cent of the cost, never mind their caste or tribal status (UNICEF 1993). The emphasis in future was to be on the promotion of a range of design and price options for household toilets, and on mobilizing communities behind sanitation goals and decentralizing decision-making. The success in Medinipur helped pave the way for the wholehearted commitment, at least at the policy level, for a change of approach. Under the TSC, which is being carried out in 27 states, 3.3 million household toilets, 1700 sanitary complexes for women, and 41,000 school sanitation blocks in schools have since been built—a marked improvement on the results of previous sanitation programmes.

The lesson was that sanitation, in the form of toilets, was not a development which could be delivered wholesale to the people. But that where communities were mobilized and affordable options provided, latent demand for sanitation did exist and progress could be made. In some districts of some states, significant change was in the air.

Nothing could provide such a stark contrast to the water-rich environment of West Bengal than the water-starved environment of Rajasthan. With 56 million people, Rajasthan contains 5 per cent of India's population, but they have to manage on 1 per cent of the country's water. Drought in Rajasthan is more common than its absence. There have been droughts in 15 of the last 17 years. During the monsoon of 2002, the fifth successive year of drought, rainfall was only 30–40 per cent of normal. The farmers of this beautiful but inhospitable terrain have well-tried ways of tiding themselves through hard times, but in the autumn of 2002, the water crisis was uppermost in people's minds. So it was not surprising to find that there was a distinct lack of interest in sanitation at that time, all energy being reserved for water conservation.

If that was the case for adults, it was not so for children. At the upper primary school in Malikpur village, Malpura block, in Tonk district, a young boy from Class 8 gave a detailed explanation of the duties of the school's 30 sanitation scouts. On Monday, his team—the Tigers—inspect everyone's nails to see that they are properly clipped. On Tuesday, the Lions check the school premises to see that they are properly swept. On Wednesday, the Peacocks (whose leader is a girl) monitor 'trees and plantations'. The Peacocks are doing a good job. In the centre of the school compound there is a large square of grass—a luxury watered by the school pump's waste output—fringed with colourful plants and flowers. On Thursday, the Leopards see that solid waste is properly disposed into the garbage pit. Friday is Scouts' day, so the teams have a day off. As for cleaning the school toilets and the handpump area, 'everyone is responsible'. The sanitation scouts have a song which they sing round the village. The song promotes washing of the hands, and the use of a long-handled ladle for dipping into the household water pot to keep its contents safe from germs, 'price only ten rupees'.

This was a display of school sanitation with a vengeance. Woe betide any student who transgressed the sanitation code. Posters and charts adorned every school classroom, citing the virtues attached to cleanliness in all religions. In the corner of the main yard stood the handpump, and the block of girls' and boys' urinals and latrines. In the corner of one classroom a model of a check-dam had been constructed, and its genuine run-off was causing a small bed of greenery to sprout.

The School Management Committee for the programme has assembled cross-legged on the school verandah for their monthly meeting, the men resplendent in Rajasthani turbans, the women hiding their eyes with the edges of their saris. The chairman is the local sarpanch, the secretary the headmaster, and the membership includes other 'eminent personalities' of Malikpur village. Their presence illustrates the school's commitment to becoming a brand leader in sanitation. Its example aims to inspire pride among parents throughout the locality in a safe and pleasant living environment. How many students use a toilet at home? Hands shoot up, but they are still a minority. However, most of the rest will surely want a toilet when their own turn to be householders comes around.

School sanitation is only one part of the support for water, hygiene, and environmental sanitation provided by Unicef as part of a 'total package' in Tonk and two other districts of Rajasthan where social indicators are low. In Tonk, the young child mortality rate is 149 per 1000 live births, compared to 115 in Rajasthan as a whole. Life expectancy is one of the lowest in the state at 49 years. Rates of diarrhoeal disease and nutritional deprivation are high, and attendance of girls in school relatively low (female literacy is 32 per cent, compared to 71 per cent male). Sanitation in the schools participating in the programme is given a high degree of importance. In this kind of social and geographical setting, so different from the hugger-mugger living of West Bengal, the spread-out character of villages in the semi-desert environment, the hardships people face, and the difficulty of changing deeply entrenched cultural traditions, means that the school can be a hub from which new ideas can spread. A show-piece environment, not in terms of fancy buildings but in making the best of what they have, is important in conferring prestige. The energy and motivation of the staff to improve the school and give the students pride and a sense of achievement in what they do there enhances their influence in the wider community.

Take the school at Durgapura in Deoli block. This, too, has undergone improvements since 2000. The school used to be situated right next to a burning *ghat* (cremation place). The students used the ghat if they wanted to go to the toilet, which many hated doing. So the ghat was shifted, the compound cleared up, and toilet blocks for girls and boys constructed. The School Management Committee managed

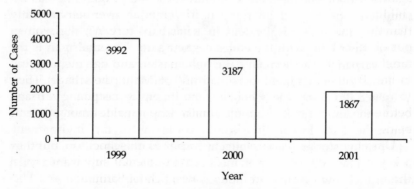

FIGURE 4.4: Impact of the Integrated Water and Sanitation Project, Tonk district, Rajasthan: decline in morbidity resulting from diarrhoea

Source: Health Centres/Sub Centres, Medical Deptt., Tonk Distt., Rajasthan.

to come up with some funds to pay for extra teachers and sanitation blocks. In 2000, 134 children were enrolled. By 2002, the numbers had risen to 204. As at Malikpur, there are daily sanitation duties. Once a month, a team of children goes around the village to talk to householders. There is a register of the 80 families with children in the school. By October 2002, 100 per cent were 'practising hand-washing', 90 per cent were using a long-handled ladle, and 27 had installed toilets. There are maps which the children have produced, pinned on the headmaster's office wall, showing every house in the village, whether it contains school-aged children and if they are attending, and whether the household has a ladle and toilet.

This should over time produce other social gains. Girls in Rajasthan frequently drop out of school at puberty. The provision of separate sanitary facilities for girls removes one excuse of parents for with-drawing them on grounds that their purity and modesty may be compromised. This is a part of the world where the age of marriage for girls is very low. sixty per cent are married by age 15. A few girl students bear the red blaze (vermilion powder) at their hair parting that shows they have already been formally wed. Retaining girls in school improves not only women's status but their future chances of having smaller families and raising healthy children of their own. Without education, girls settle into married life and become mothers in their teens, perpetuating in their turn the pattern of discrimination against girls from the earliest days of their lives. Poor care in early childhood means that Rajasthani girls have a lower survival rate than their peers almost anywhere else in the country. When girls drop out of school at Malikpur and Durgapura, staff members visit the families and dissuade them from reducing their daughters' chances in life. Thus the provision of water and toilets in the schools helps to retain girl students, which in turn helps to postpone marriage before the legal age of 18 and gradually raise female status and well-being.

Unicef currently places a strong emphasis on school sanitation as a key component of its water and environmental support strategy throughout the country. In many states, School Sanitation and Hygiene Education (SSHE) is now regarded as the 'entry point' for the promotion of sanitation in the community. Unlike in West Bengal, where the production and sale of toilets to individual households is driving community health and cleanliness forward, in some states, including Karnataka, schools have been allotted this role. Here too, the provision of water and sanitation facilities is linked with overall improvement: better classrooms, more teachers, new interactive methodologies, different textbooks, boundary walls to keep grazing livestock and rubbish out and environmental improvement in. There is strong political and administrative support for what is seen as building a people's movement for a cleaner way of life: 'The schools are places where our future administrators are being raised,' proclaims the Secretary of Rural Development and Panchayati Raj, Kaushik Mukherjee.

In 2000–2, the programme operated in six districts, of which Mysore was the pace-setter. One of the three blocks or *taluks* in Mysore earmarked for improvements in 300 schools was K. R. Nagar taluk. Here, 110 of the 235 gram panchayats applied for the programme: a clear indication of its popularity. The gram panchayat has to find 50 per cent of the costs of construction, and this is matched by 50 per cent provided by Unicef. All the training, orientation, and the IEC—the 'software'—is also funded by Unicef. The motivation of teachers to undertake the programme is critical: 'You must have a spark,' says Manoj Kumar, the project coordinator attached to the *zilla* (district) panchayat headquarters in Mysore.

At Doddekoppalu primary school, that spark is provided by the assistant teacher, Naveen Kumar. He exudes enthusiasm for the transformation of the school he has promoted under the School Water and Sanitation towards Health and Hygiene programme. This is not a large school—150 students only—and is modestly housed. The compound used to be open to the rest of the community, a place where people came to relax, drink, smoke, and chat, and also, regrettably, to defecate. But ever since the area has been enclosed by a boundary wall, everything has changed. The premises (including the toilets) are clean and freshly painted, and the yard is a riot of colour, with beautifully laid tiny flower beds and paths in the shape of the Indian flag, herbs, shrubs, even a rock garden. 'This is not a school, it is a heaven,' says Naveen Kumar, and everyone smiles with pleasure and applauds.

A cabinet elected from among the students plays an important part in helping to keep the premises clean. There is a child minister for culture, minister for finance, minister of the environment, and so on. Each of the nine cabinet members is attached to a teacher. Each class collects weekly contributions from its members to pay for cleanliness extras—'soaps and brushes', for example. A kitty of around Rs 60 or 80 is contributed per month by each class, and pooled with the proceeds from others. Each has a passbook with all the entries carefully listed. So enthused by the programme have they become that the students are prepared to give up their pocket money—given to them by their parents to buy sweets with—to the school sanitation effort.

The success of the programme and its popularity with the gram panchayats has attracted considerable official attention. The then Chief Minister, S. M. Krishna, formally requested Unicef to extend the programme to other districts and disadvantaged taluks, and there is a constant stream of letters from politicians anxious that it be taken up in their constituencies. The potential for increasing their votes, come election time, by their association with a highly visible programme is not lost on them. Other external support agencies have also shown interest in contributing funds. This success, shown by the pressures and demands coming both from on high, and from communities themselves, is gratifying to Unicef, which spent many years promoting the programme against initial inertia. But sorting out now how to go from demonstration of an effective model to statewide implementation is not easy. Often when programmes are taken up on a significant scale by the government, never mind the commitment on paper to the need for 'software', social mobilization, IEC, and behavioural change, what happens is a bonanza of construction. Instead of a transformation of mindsets and sanitary mores, the programme could easily become a pretext for what Mukherjee himself described as COW: 'contractor-oriented work'.

In the case of school sanitation in Karnataka, there is a very real problem. This is a state with a rich agricultural output, a high proportion of which is rice. Land, especially good and well-watered land, is at a premium. A number of schools were established on a piece of land given for the purpose by a landowner many decades ago. This land was regarded as waste, which is why it became a dumping ground, a place where animals might stray and which no one bothered to look after, often with the school buildings huddled in a corner.

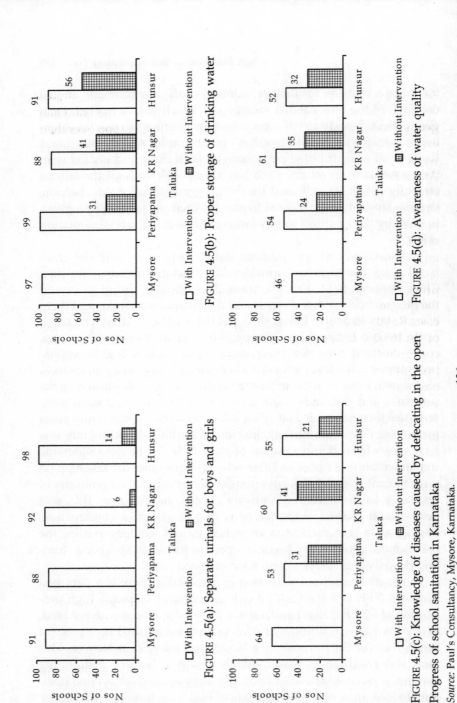

FIGURE 4.5(a): Separate urinals for boys and girls

FIGURE 4.5(b): Proper storage of drinking water

FIGURE 4.5(c): Knowledge of diseases caused by defecating in the open

FIGURE 4.5(d): Awareness of water quality

Progress of school sanitation in Karnataka

Source: Paul's Consultancy, Mysore, Karnataka.

As incomes and land prices have risen, some descendents of the original 'donors' have tried to dispute the perpetuity of the grant and get the land restored to them. Many schools are in litigation over their exact perimeters. No wonder schools want good, solid boundary walls. Besides, every decent building in the area is similarly equipped. And it is true that, without some protection from unwelcome visitors, stray animals, litter, and the rest, no school compound can be safely smartened up nor can head teachers go to bed free of worry that the new water taps and toilets would not have been vandalized by the morning. So sturdy boundary walls are now accepted as an essential ingredient of the school sanitation package.

Boundary walls, however, are expensive. Described in Unicef's programme literature as 'protection to Watsan facilities' (Zilla Panchayat, Mysore 2002), they are by far the costliest item. Each wall costs Rs 90,000 (around US$ 2000) out of a total infrastructure budget of Rs 165,000 for each school. Training, posters, and other software cost only Rs 29,000. The wall is often more expensive and strongly built than the whole of the rest of the school. Up to now, Unicef has paid half the costs of boundary walls. So Unicef has effectively been subsidizing the construction budget of the Ministry of Education. Strictly speaking, the ministry is responsible for all school buildings including classrooms, boundary walls, water points, and toilets. But education budgets are under pressure after years of erosion.

Unicef funds are not normally used for the construction of school buildings, but for handpumps, pipes, water tanks, and toilets it makes an exception. Now the definition of sanitation has been stretched to include 'protection to facilities' because, in the unanimous view of teachers, parents, communities, and government officials, boundary walls, not only in Karnataka but in Andhra Pradesh and in some other states, are critical. But having made this discovery and having established the model, the state has to be encouraged to take over this responsibility, perhaps with contributions from the community as well. In some environments, fencing may provide a less costly alternative. In Assam, for example, almost all schools are surrounded by bamboo picket fences, reinforced by cacti or other fast-growing shrubs. The selection of fencing material is bound to be a local decision, depending on what is regarded as appropriate and acceptable in different settings. But this question is one in which Unicef should only be peripherally engaged. Unicef is not a construction agency, nor does it undertake mass implementation of government programmes. So on the threshold of the kind of expansion and scaling up that some programmes only dream of, there is still a hurdle to overcome.

Can the school sanitation programmes, in Karnataka, Andhra Pradesh, and elsewhere, attract the large-scale political, administrative, and budgetary support that rural water supplies managed to capture in an earlier era? That is now the big question. Many committed sanitation officers, in the government and in Unicef, believe that it can. They envisage schools as centres of opportunity to transform environmental conditions in communities and reduce sanitation-related disease on a significant scale. But for that to happen, the whole strategy will have to be adopted lock, stock, and barrel by the central and state governments, not just parts of it, and significant budgetary resources will have to be pushed in its direction. This will require extra money for educational infrastructure generally. It will also require real commitment, which has often proved elusive in many state-run programmes, to software: training, orientation, and IEC. There is still a long way to go.

REFERENCES

Black, Maggie (1990), *From Handpumps to Health*, UNICEF, New York.
Dixit Annapoorna (1998), 'Table of developments in the sanitation programme', in paper contributed to the 1998 National Seminar on Rural Sanitation.
Drinking Water, sanitation and hygiene in India, NSS 54th Round (January–June 1998), National Sample Survey Organisation, Department of Statistics, Government of India, July 1999. Source of estimate, NSS, 38th round, January–December 1983.
Government of India (2000), *Multiple Indicator Cluster Survey (MICS-2000)*, Department of Women and Child Development, GOI, and UNICEF.
———— (1991), *'Rural Sanitary Marts'*, UNICEF and GOI Water Mission.
———— (2000), *Mid-term Review of the Ninth Plan*, Study Reports by the GOI, quoted by N. C. Saxena in his paper on 'School Sanitation', p. 7.
Government of West Bengal (2002), 'Rural Sanitation Programme and Sanitary Mart', leaflet issued by the Panchayat and Rural Development Department, West Bengal.
Nadkarni, Manoj (2001), 'Drowning in Human Excreta', *Down To Earth* magazine, Vol. 10, No. 19, CSE, New Delhi, 28 February.

Naipaul, V. S. (1964), *An Area of Darkness*, Penguin, London.

Ramakrishna Mission (1994), *Child Welfare and Development*, a brief report, Ramakrishna Mission Lokasiksha Parishad, Narendrapur, West Bengal.

Samanta, B. B. (1997), 'Role of UNICEF in transfer of technology', Orissa, quoted in UNICEF's *Learning from Experience*.

Saxena, N. C. (2003), *School Sanitation Programme*—a report to UNICEF, Delhi, page 6.

UNICEF (2001), 'Evaluation of Rural Sanitary Marts and Production Centres', Select Case Studies, prepared by TARU Leading Edge for UNICEF, New Delhi, February.

———— (2000), *Learning from Experience: Evaluation of UNICEF's Water and Environmental Sanitation Programme in India, 1966–98*, Evaluation Office, UNICEF, New York.

———— (1996), *Sanitation—The Medinipur Story*, UNICEF, Kolkata.

———— (1994), *Project Profile: Intensive Sanitation Project, Medinipur District*, West Bengal, UNICEF Watsan, New Delhi.

———— (1994), *Sanitation for Better Health*, UNICEF, New Delhi.

Wan, Philip (1988), *Sanitation Programme, India, 1982–8: Reflections*, UNICEF, New Delhi, April.

Water Aid (1996), 'Thirsty Cities: Water, sanitation and the urban poor', original source, *The Economist*, 7 October 1994.

WHO (1992), *Our Planet, Our Health*, report of the WHO Commission on Health and Environment, WHO, Geneva.

Winblad, Uno (ed.) (1998), *Ecological sanitation*, SIDA, Stockholm.

Zilla Panchayat, Mysore (2002), 'Action Plan for School Sanitation in Mysore District 2001–2'.

5

Tackling Disease:
Guinea Worm and Diarrhoea

Dracunculiasis is unique among water-associated diseases in being the only one uncompromisingly associated with drinking unsafe water and having nothing to do with excreta, dirt, or germs. The only way to contract dracunculiasis or guinea worm[1] is to drink water that contains a tiny water flea or cyclops containing the parasite *dracunculus medinensis*. The name of the parasite and of the disease comes from a Greek word meaning 'very harsh, inhumanly severe or cruel'. The pain this parasitic worm causes its human host is acute, lasts for several weeks, often causes fever, nausea and vomiting, and can lead to permanent impairment.

Dracunculiasis is easily avoided if one understands how it works. The pinhead-sized cyclops breeds in standing water—wells, tanks, rice fields, irrigation channels, turbid streams, and stagnant ponds, and absorbs the dracunculus parasite, becoming its host. When imbibed, the cyclops dies in the human stomach, but the parasite migrates to subcutaneous tissues. Over several months, it matures into a worm of up to a metre in length. After around ten months, this milky-white, semi-transparent living string begins to emerge from the body, sometimes from the trunk, often from a limb or extremity, even from the sole of the foot. This creates an excruciatingly painful

[1] The common name for the disease was coined by sixteenth century European travellers in parts of West Africa bordering the Gulf of Guinea, who recorded seeing cases. West Africa remains the last major repository of guinea worm.

burning ulcer. So the patient seeks relief by immersing the ulcer in water. Unfortunately, this allows the worm to release hundreds of thousands of new parasite larvae into the water which in turn become absorbed by the cyclops, thus completing the cycle which will bring infection to next year's sufferers.

This detested parasite has been known in India since prehistoric times. The *Rig Veda*, believed to have been written around 1200 BC, includes a plea, in a poem by the sage Vasishtha: 'Let not the sinuous worm strike me nor wound my foot'. The life cycle of the dracunculiasis parasite was first described in the late nineteenth century, but the scientific explanation for what in much of the world was known as 'the fiery serpent' has inevitably taken a long time to penetrate to the remote, backward, and largely illiterate communities where such afflictions of poverty flourish. In Jaisalmer, Rajasthan, the grave of Ram Devra, a fifteenth century sage, attracts thousands of devotees. Local legend decreed that *naru* (guinea worm) was a manifestation of Ram Devra's wrath. Patients visiting the shrine were said to have been miraculously cured (UNICEF 2003). In such remote corners of harsh terrain it is not easy to break centuries-old superstitions that pain, and its cure, are in the hands of gods, saints, and sinners.

In 1983, India became the first country in the world to launch a national programme aimed at total guinea worm eradication. The idea had been mooted in 1979, and the country's epidemiological experts and health infrastructure had been gearing up for the past few years. At that time, there were thought to be around 40,000 cases annually in seven endemic states: Andhra Pradesh, Gujarat, Madhya Pradesh, Maharashtra, Karnataka, Rajasthan, and Tamil Nadu. The total population at risk was thought to be five million (WHO and UNICEF 2001). Almost all of the victims were in poor rural communities, many of them tribal. The period when the worm was most likely to start its dreadful trajectory through the skin was in the pre-monsoon summer months when agricultural activity was at its height. Men were more frequently struck down than women, with older children of both sexes also affected. A patient spent an average of 40 days incapacitated by painful ulcers. This meant that he or she was unable to work in the fields, undertake household or domestic responsibilities, or go to school. The disease thus plunged many marginalized families deeper into poverty. Since there was ignorance and superstition about the affliction, no effort was made to protect water bodies from infestation, and victims often re-infected themselves as they took refuge in the cooling effects of the streams and ponds to soothe their burning sores, and drank the contaminated water.

Although in principle the eradication of guinea worm appeared simple—the requirement being to break the cycle of transmission by preventing new larvae entering water bodies—in practice any campaign to eradicate a disease is highly complex. Such a campaign requires planning and implementation of military precision, and the involvement of a large interconnected web of local teams and supervisory layers, through whose mesh no worm or sufferer can be allowed to fall. With guinea worm, there is no drug or vaccine. Everything depends on prevention. So putting messages across to people rather than lining them up for a medical fix was the only campaign tool. Every last case had to be hunted down with pinpoint accuracy and every last potential breeding ground for infection tackled. So case detection was a central feature of the eradication strategy, to be followed by various kinds of intervention. These included treating water sources with insecticide, protecting them from human contact or closing them off and providing handpump–boreholes instead, educating people to filter their drinking water, dealing effectively with patients, and raising community awareness.

In the early 1980s, a series of surveys were conducted to identify affected households and water sources. Each time a new survey was undertaken, the number of affected villages and people potentially at risk rose considerably. This is normal in situations where, previously, data has been sketchy. It reflects increasing efficiency in survey methods, including selecting the time of year in which prevalence is likely to be the highest. Between May–June 1981 and June 1982 the number of affected villages rose from 7533 to 11,736, and the population at risk rose from 5.9 million to 12.6 million (UNICEF 2003). The worst-affected state was Rajasthan, with over 6000 villages and nearly 15,000 patients. The good news was that, by 1982, Tamil Nadu had already succeeded in eliminating the disease.

Once the eradication programme was launched in 1983, the case detection strategy used both the regular health infrastructure reporting system, and special case-searching operations. If an area was known to be endemic, three such operations were launched every year, two before the main rainy season, and one in November/ December when many areas have a second less important rainy season. In a coordinated campaign, health workers and supervisors from the Primary Health Centres (PHCs) set out to visit an allotted number of villages and habitations, undertaking a door-to-door inquiry for guinea worm patients. Surveillance was also conducted on a continuous basis through the regular health services and outreach workers: multi-purpose health workers, anganwadi workers, and auxiliary nurse-midwives (ANMs). Monthly reports were sent up the chain from PHCs to district and state authorities, to be processed by the National Institute of Communicable Diseases (NICD) in Delhi, the nerve centre of the campaign. The NICD deployed 12 epidemiological surveillance teams, whose job it was to go to endemic areas, track the guinea worm situation in detail, and conduct surprise checks on programme activities.

The other key area was to identify all the safe and unsafe community drinking water sources. These were mapped in endemic areas, and those liable to guinea worm infestation were then protected or converted. In many water-scarce and guinea worm infected parts of the country, people drew their water from traditional wells, tanks, and ponds. Patients with worms protruding from their limbs had to be prevented from entering the water source. The solution was to fence ponds, and erect a device allowing water to be drawn by bucket, rope, and pulley. Occasionally, in the last stages of eradication from a given area, guards were employed to keep vigil over the water source against intruders who might be suffering from guinea worm.

In Rajasthan, step-wells with wide circumference were an important traditional water source. The water carrier descended via steps into the well to fill her water container. In the grandest wells, which were effectively large walled tanks, she went down flights of stairs as at a bathing ghat. She walked into the water to lower her jar or pot, lingering to wash, and often to do her laundry squatting on the steps. Access on foot to the wells had to be cut off, by blocking off stairs and chipping away steps. Then a pulley would be set up and the step-well converted to a draw-well. None of this could be done without community consultation. Traditional wells could not be closed or altered out of recognition in an arbitrary fashion. But villagers were less resistant to giving up their old ways of bathing, laundry, and water collection if they could be convinced that it would release them from the scourge of guinea worm.

As well as reducing access to water sources, efforts were made to kill the cyclops in the infested water bodies. Chemical treatment using a safe insecticide—Temephos—was initiated for all open sources in infected areas. PHC workers were trained to calculate the required dosage for a water body and to apply it safely. Cyclops densities were also monitored regularly by district, state, and national supervising officers, and forwarded up the line to the NICD, which evaluated the treatment and coverage of sources. Applications of Temephos were gradually increased, and from 1989, occurred once a month from February through June, and every two months from July through December.

As a further precaution, people were encouraged to filter their drinking water in the household to be certain of not ingesting any stray cyclops. Special filters to be fitted over the water pot or jar were supplied to households in infected areas. These were double-sided, white on one side, coloured on the other, so that the clean and the contaminated sides could not get confused. Where they were not available, village women were taught to use a double layer of cloth. Some continued to do this long after guinea worm had ceased to be a threat in their location. Meanwhile, guinea worm victims were given treatment. The age-old method was to take hold of the worm after it began to emerge, and pull it carefully out centimetre by centimetre, day by day, and roll it around a stick. It was important for the worm not to break, because if a length of worm was left in the body, it could cause serious allergic reactions. PHC workers were trained to roll the emerging worm onto a sterilized bandage. The process would still take several days. But the ulcer would at least be bandaged, helping it to heal more quickly and preventing the discharge of larvae into the open. If the worm broke, they were given tetanus toxoid and extra care.

Although the programme began to show results quite quickly, the campaign faced a major difficulty in transforming guinea worm reduction into full-scale eradication. To deliver 100 per cent case identification and containment is the largest problem faced by any disease eradication programme. Following the establishment of the

National Drinking Water Mission in 1986, with a special sub-mission for guinea worm, the campaign became more focused and dynamic. The rapid increase in the spread of handpump–borehole drinking water supplies during the 1980s made an important contribution— both central and state governments put a priority on installing handpump supplies or piped water systems in guinea worm affected communities. But it became clear by the late 1980s that there had been insufficient emphasis on 'software' and on mobilizing communities to participate. In any disease eradication campaign which cannot deploy a technical fix such as a drug or vaccination, in the end, it is the change in people's knowledge and consequent behaviour which determines the success of the campaign.

The challenge was at its most acute in the deeply traditional and distant parts of Rajasthan. In 1984, over half the guinea worm cases in India came from this one state, and 70 per cent of the patients in Rajasthan lived in the four southern districts. What happened in those would be critical to guinea worm eradication not only in the state but in the country. So, with support from Unicef and funds from Sweden, a special initiative was begun there in 1986. This project came to be known as the Sanitation, Water, and Community Health project, or SWACH which evokes the Hindi word, 'swachchh', mean-ing 'clean'. This project was to have a major impact not only on India's banishment of guinea worm, but on the design of sanitation and water programmes elsewhere in the country.

The countryside in southern Rajasthan undulates gently, scrub alter-nating with stretches of reddish brown earth, out of which erupts the occasional rocky hill, sometimes topped by an ancient fort or battle-ment. Although this is not true desert country as in the north and west, it is dry and dusty, and in the season before the monsoon, the temperature reaches up into the high 40s. Hot winds blow across the open landscape with the force of a blast furnace, tearing at the sparse vegetation and making clothing flap and billow like tortured sails. Earthen villages merge into the landscape, and out where there is something resembling pasture, the bright red turbans of the herds-men draw the eye like distant beacons. Most of the people in this area are adivasis. Dominated for centuries by the *Rajputs*—warrior rulers who amassed fortunes from the movement of silk, spices, and precious stones along lucrative East–West trade routes—this is a society of rigid tradition, still part mediaeval in its outlook. *'Delhi door ast'*, is the saying: 'Delhi is far away'.

The SWACH project, with its headquarters in the princely city of Udaipur, started life in 1986 with a mission to eradicate guinea worm from southern Rajasthan. With a population of 1.8 million, the two districts of Banswara and neighbouring Dungarpur contained 27 per cent of the 6100 affected villages in the state and were the first to be targeted. Once the project had developed and refined its methodology for communications outreach and case detection, Udaipur district—with 6152 cases, the largest number in any district in the country—and Rajsamand were added.

From the outset the SWACH project managers—dynamic and visionary Indian Administrative Service (IAS) personnel—placed an emphasis on health education and environmental sanitation connected to guinea worm transmission. Before long the project expanded to embrace a wide range of health, water supply, sanitation, and disease-prevention concerns. In Rajasthan infant mortality rates were 300–400 per thousand live births, compared to 110 for the country as a whole, and they were at their lowest in the southern districts. Almost all those affected by guinea worm lived remote, harsh, and isolated lives, and in common with many adivasi communities, were so far off the beaten track that they had been overlooked by the emissaries of the modern world. The campaign would bring many of the state's poorest people into first time contact with outsiders proposing changes in habits practised since time immemorial. Many had never seen a doctor or a midwife or been to a hospital, and found the things that went on there incomprehensible and terrifying. They would be introduced to a scientific explanation for the disease, including the 'fiery serpent' that had cursed their lives for so long, and a medical response. If persuasive, the campaign would not only deal with the parasite, but provide an entry point for better hygiene practices and maternal and child health generally.

The innovatory character of SWACH's methodology was to focus on education and mobilization, not just on breaking transmission by putting physical barriers between people and the guinea worm cyclops. This required the training of all personnel undertaking technical, social, and health-related activities in communication techniques of many kinds. Chiefly, they had to be able to put across the essential programme messages in a village setting. Every person employed in the programme, whether a driller or a social communicator, was expected to be fully conversant with its overall aims and activities. Since the target of so many of the project's benefits and exhortations to adopt new ideas were women, half the project's staff were women. District training teams were established with members from the education and health departments, NGOs, and social services. Their task was to develop village cadres who would weave into the fabric of community life a new code of behaviour regarding water and environmental sanitation.

Action at the community level with participation from everyone was at the heart of the strategy. One key activity was the 'village contact drive', so called because of its purpose—to establish a link between rural communities and officials or project people. The contact drive gave villagers a chance to hear what the project had to offer, and to respond. The exercise also gave the project staff a chance to learn about the villages and collect basic data. In the initial stages, the target villages were selected by examining medical records to find out which were the most affected by dracunculiasis. Over time, almost all villages were contacted, whether they were recorded as having guinea worm patients or not. During the first three years, over 3000 villages were contacted during the course of two contact drives.

The drives were conducted by teams of young people, usually two women and three men. One was the *gram sevak* (village helper), the lowest level government development worker. The other team members were selected by soliciting names from all possible sources: schools, panchayats, NGOs, local clubs. All the teams took part in a four-day training camp, during which they assumed different functions, planned their itinerary, organized their work schedule, and shared out the various tasks. The actual drive took place over a period of 15 days, and was a cross between a *yatra* (a march), a pilgrimage, and a hiking tour through the countryside. These were large events. For the second village contact drive in Banswara, 123 teams—800 people altogether—took to the roads at the height of the hot season. Few of the young women had ever before spent the night elsewhere than in their own homes.

Each group walked to the closest village on their list, carrying bed rolls, a small daily allowance for food, puppets, paint, brushes, and a repertoire of slogans and songs. On arrival, they told the local sarpanch the purpose of their visit and asked permission to explain to everyone the nature of the project and collect information. The women team members gathered all the village women together, discussed where they collected their water, how many step-wells and other water sources they had, and whether there were cases of dracunculiasis. They would explain all about the life cycle of the guinea worm, and give out double-sided filter cloths. They canvassed views about protecting the step-wells, asked where handpump–boreholes might be put, wrote slogans on the walls of schools and buildings, and distributed posters. In the evenings, they put on shows using puppet plays, songs, and traditional stories. They then slept in the village before proceeding on to the next destination to repeat the procedure. As time went on, the contact drives adapted and refined their activities, testing out different messages and songs, and leaving their routes flexible so as to make detours if they heard of a nearby guinea worm case.

The other cadre of personnel the project introduced was village-level animators. The help of the Women's Development Department was sought in selecting and vetting candidates: young women volunteers who would be given a brief training and a small financial incentive to promote healthy behaviour in the community on a day-to-day basis. Unlike many health outreach programmes using volunteers, where candidates are usually selected by the communities and may not be highly motivated, the project was determined to find women keen to perform well, and strong on communication skills. They needed stamina and perseverance for periods in between the contact drives when community interest flagged, and members were less responsive to the need to construct soak-pits for drainage, use filters for drinking water, and avoid contamination of water sources with guinea worm larvae. To avoid a high drop-out level, hundreds more women candidates attended training camps and were observed during field activities before selection. Eventually, the project aimed to leave in place 360 social animators altogether.

The animators were not left to function independently—a typical failing of projects trying to be 'community-based'. They were grouped into fours and fives, with one group acting as supervisor of the rest. They concentrated their activity in two or three villages at a time, observing how people were using the step-wells or newly-installed handpumps, and whether they were filtering their drinking water. They were encouraged to link up with NGOs working in the area, and develop ties with anganwadi and other community development workers. For their duties, they received a small stipend every month for the duration of the project. Although this meant that the incentive to carry out their activities in a regular way would end within a few years, at least the guinea worm might be banished by then, and there was a chance that some of the animators might be sufficiently inspired by their experience to start up their own NGOs or get involved with other programmes.

Another special duty of the animators was to keep up to date with cases of dracunculiasis. The SWACH project felt that treatment for patients must be an essential part of its work. It employed Dr Sharma, an *ayurvedic* doctor, who had specialized in the extraction of guinea worms for over 30 years. His technique could be used when the worm had been detected under the skin but had yet to make its appearance. The surgical extraction of guinea worms by ayurvedic doctors was, however, controversial. The NICD, the institute in overall charge of the campaign, was deeply resistant to the idea of traditional medical practitioners undertaking surgical procedures outside the sterile conditions of a hospital. However, the method was approved by a team of senior surgeons from the All India Institute of Medical Sciences (AIIMS), who described it as 'neat, quick, with minimum trauma to the patient. None of the patients complained of any pain during the procedure' (UNICEF 2003). Other doctors were later trained in the same extraction method.

When an extraction camp was due to be held in their locality, the animators would send word to all the patients in the area. In the past, few patients presented themselves for surgical extractions since they were too poor to pay the fee charged by private doctors. Now the treatment was free. Apart from relieving patients' suffering and enabling them to avoid weeks of pain, the operations, especially when performed before the worm's emergence, reduced dramatically the risks of further transmission of the disease. The camps also provided an advocacy opportunity for other health-protection measures, and helped monitor caseloads and progress towards eradication. Over the course of a few years, thousands of guinea worms were removed by this form of surgery in hundreds of camps. At a later stage, newly-trained ayurvedic guinea worm surgeons were posted to field hospitals at the block level, where they could handle up to 10 resident patients at any one time. However, the camps were more popular than hospitals. Many tribal people were fearful of staying in such a place for an 'operation'. The very last known guinea worm patient caused consternation when he ran away from hospital in the night, terrified of staying in such a place by himself. A major search had to be conducted to find and reassure him.

At the same time, there was a concerted effort to control access to step-wells and speed up the delivery of new boreholes and hand-pumps. Over 2600 step-wells were protected or converted within two years. A further 4000 handpump–boreholes were added to the 9000 already installed in the two districts by the PHED, using its own equipment and extra drilling rigs provided by Unicef (UNICEF 2003). Many of the new handpumps were close to step-wells so that fully protected water sources could be found in a familiar location. The project was, therefore, strong not only on social mobilization and health education, but on technology too. The hardware side of the project was far more comprehensive than was usual in rural water supplies provision, and took into account all water-related household activity, not merely the need for drinking water.

In some locations there was a refinement to the standard site design. An extra platform apron was added so that run-off from the handpump could flow into a trough for watering cattle and other livestock. This made it easier to keep the area immediately around the handpump clean from contamination by their excreta. Waste water was also led away to irrigate fruit trees or a kitchen garden. Water was so scarce in Rajasthan that not a drop could be wasted. Special raised slabs were also built for washing clothes—so integral an activity at traditional step-wells that it was essential to provide an alternative. SWACH also pioneered the training of women mechanics to keep the new handpumps operational. These were in far-flung areas where existing maintenance systems were not working

One of the earliest drilling rigs in India, the Halco Tiger,
Andhra Pradesh, 1969

The India MK II handpump, Tamil Nadu

Low-cost toilet, Jharkhand

Transporting toilet pans in West Bengal

Women employees at a production centre in Medinipur District,
West Bengal

The Tara Handpump in West Bengal

Handpump maintenance, an income-generating activity for rural women

The India MK II
in Jharkhand

Social mobilization of communities for water supply management and
public health, Rajasthan

Monitoring health and sanitation interventions, Tonk District, Rajasthan

Hand-washing and toilet use, school sanitation programme,
Tonk District, Rajasthan

Ayurvedic pioneer in the surgical extraction of guinea worn, Dr. Bhanwarlal
Sharma operates on a Rajasthani youth

Child-crippled with fluorosis

Fluoride removal filter in a Rajasthani household

Roof-top rainwater harvesting at a school in Aurangabad District, Maharashtra

Check-dam, Aurangabad District, Maharashtra

and back-up teams from the PHED were not at constant beck and call. Whenever there was a breakdown, people were forced back onto open sources and the chances of re-infecting them with larvae quickly rose. So there was a real incentive to try out community maintenance of handpumps. That the SWACH project selected women for this task in such a deeply conservative environment was thought extraordinary by officials and communities alike. However, it paid off and was later replicated in other drinking water programmes using India Mark II and Mark III handpumps and Tara pumps (see Chapter 6).

In the early 1990s, a big international push was given to the goal of eradicating dracunculiasis globally. WHO set its sights on 1995 as the target date, but, more realistically, the 1990 UN Children's Summit set the goal of full eradication by the end of the millennium. India stepped up its national efforts as well. Maharashtra and Gujarat witnessed their last cases in 1991. By this stage, SWACH had guinea worm on the run in the four southern districts of Rajasthan, but the programme continued in existence until 1994 to mop up the final black spots. By then, guinea worm had been fully eradicated from Banswara. In Dungarpur, the caseload dropped to five in 1993 and in Udaipur, to four. In 1995, no further cases from these districts were reported (UNICEF 2003).

By now, attention had begun to shift to the other nine affected districts in Rajasthan, which between them were home to 600 affected communities. In 1992, the Rajasthan Integrated Guinea worm Eradication Programme (RIGEP) was launched to replicate the successful SWACH strategy in the districts of western Rajasthan where cases were still being reported. New teams of social animators were trained, and an extra cadre created—scouts, to conduct house-to-house surveys. Within a year of the launch of the networks, the management team was able to respond within 24 hours of a case being reported to the district headquarters and despatch a medical team to the village in question.

In the far west of Rajasthan, the Thar desert reigns and the water table is very low. Here, life is nomadic and step-wells are few and far between. Most communities rely on *nadis* or ponds. These are natural surface depressions that receive rainwater from one or more directions. Some nadis have stone walls on one or two sides to capture and retain extra water. In normal years, they may hold water for up to eight months, while a few contain water all year round. Each desert village has one or more nadis and surveys found nearly 4000 of them in the three districts of Nagaur, Barmer, and Jaisalmer. Poor maintenance and failure to de-silt had led to pollution and the presence of guinea worm, water hyacinth, and algae. Temephos disinfectant was used to spray these ponds to destroy the cyclops—unfortunately, without much effect because the water bodies were too large. So retired military men were appointed as 'watchmen' to prevent people or animals from entering them.

At the final stage of the eradication programme, an incentive system was set up to deal with outstanding cases. In Maharashtra, the prize was as high as Rs 10,000 (US$ 330). In Rajasthan, radio jingles and wall slogans proclaimed that 'every guinea worm patient will receive Rs 1000'. In addition, an informant received a fee of Rs 500 per worm, and patients received Rs 22 in compensation for lost wages for each day spent in hospital recovering from treatment. In 1994, one family from Chadi village in Jodhpur district had six guinea worms between five infected members. The head of the family, Alsa Ram, took his brood of patients into Ossiyan Sentinel Hospital. When they emerged, they were over Rs 6000 richer than when they arrived. Alsa Ram said afterwards: 'The incentive money was a windfall and certainly useful, but freedom from naru is priceless.'

During 1997, 1998, and 1999, no new cases of guinea worm were reported from Rajasthan or anywhere else in the country. A 10-member national commission of experts was appointed by the central government to decide whether the fiery serpent had definitively been banished. When they reported positively, India applied to WHO's international commission on certification to visit the country and declare it guinea worm free. A three-member team did so in 1999, toured the previously endemic states, and interviewed hundreds of health officials, surveillance personnel, and people from affected villages. India was declared free of guinea worm in 2000. Of the total costs of the campaign, the central government had contributed around US$ 10 million; the state governments, US$ 1.2 million; WHO US$ 0.7 million, and Unicef (mainly from Swedish aid for the SWACH project) US$ 17.3 million (WHO and UNICEF 2001).

SWACH was, therefore, by far the most expensive component of the overall programme to eradicate guinea worm from India—

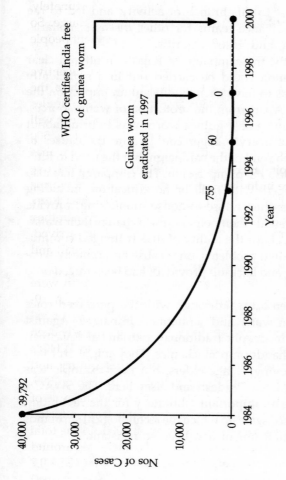

FIGURE 5.1: Decline in guinea worm cases

Source: National Institute of Communicable Diseases (NICD).

although much of the expenditure was on equipment which was later absorbed into the regular rural water supply programme. The SWACH approaches to environmental sanitation and health education had an important subsequent effect on other water, sanitation, and disease-control programmes. The emphasis on epidemiological data and surveillance gave a boost to this branch of activity, and led to the setting up of local surveillance teams for other diseases: malaria, measles, plague, cholera, and gastro-enteritis.

Even more importantly, the campaign in Rajasthan offered clear proof that health education could be carried out in a sufficiently comprehensive way so as to have such significant an impact on the prevalence of a disease. Apart from the protection of water sources, the only possible strategy against guinea worm was to bring about behavioural change. Not every villager could recite the causes of guinea worm disease at the end of the campaign. But the need to filter drinking water had made a deep impression. The campaign had also demonstrated that villagers with little or no education, including women, were able to conduct house-to-house monitoring, provide accurate reports, and persuade their neighbours to change their water-use patterns. Scepticism about the ability of disadvantaged communities to deal with their own problems, once cause and remedy were pointed out and training and back-up provided, had been forced onto the retreat.

Finally, there had been a breakthrough with the perceived roles and status of women in water and sanitation campaigns. Against difficult odds, women in deeply traditional, patriarchal Rajasthan had become practising handpump mechanics. They might still face problems from disapproving village elders, but their example was already being copied in Uttar Pradesh and elsewhere. The SWACH project turned out to be an important laboratory for the revolution in water and sanitation 'software' to which the technological supremos of the public health establishment had been so resistant.

That revolution was long overdue. In 1986, just over midway through the Water Decade and at the time when the National Drinking Water Technology Mission was being set up, a governmental working group on health education and community participation delivered a report on the necessary measures to transform the rural water supply and sanitation programme from a technical operation into a scheme for genuine social and health improvement. The report began: 'It is a truism that huge investment in water supply and sanitation without proper health education and community involvement will not have the desired results (UNICEF 1989).' This aspect of water supply had been repeatedly emphasized in report after government report ever since the Fifth Five-Year Plan, and just as consistently set aside.

The report was published by the Apex Committee responsible for guiding Water Decade activity in the country at large. It pointed out that no infrastructure or cadre existed at the state, district, or PHC level for carrying out health education. The report went into great detail about what needed to be done in terms of personnel and infrastructure, and recommended that a large allocation of resources for this purpose be included in the Seventh Plan (1987–92). In the event, the recommendations were ignored.

A Unicef consultant, reviewing the state of the art in rural water supplies and sanitation in 1989, deplored the fact that the government seemed to be withdrawing from its health education responsibilities and leaving the task to NGOs. He saw this as compromising the whole government programme, which could produce no evidence of health improvement. In fact, since 1980, water-associated disease incidence had risen, or at best remained stagnant. 'This is hardly surprising given that the rural water supply and sanitation programme is virtually void of any health education capable of influencing patterns of water utilization, collection and storage,' he concluded (UNICEF 1989).

While GOI policy remained resolutely fixed on technical solutions, and continued to value Unicef's collaboration primarily as a department for international procurement and technical advice, the trends in international thinking around basic water and sanitation services were moving in the opposite direction. Many organizations and experts involved in public health engineering were deeply concerned that large investments in water and sanitation facilities had had little impact on child or community health or on water-associated disease reduction. Since this had been the expectation, based on a simplistic reading of the story of sanitary revolution in the industrialized world, there was a movement among some public health practitioners to downgrade the importance to be attached to technical solutions—now regarded as essentially 'solved'—and look instead at social parameters. The new focus on 'software' demonstrated a professional concern with the development process—on how practices and behaviours could be changed so that life became healthier and more productive. The SWACH project was only one example among many around the world of how a technical problem of disease reduction would in the end only be solved by getting the social process right.

In Unicef at this time, the water and sanitation chiefs were all for moving in the new direction, a shift encouraged by its own major donors—particularly the Danes and the Swedes. But they were bucking a trend in the organization as a whole. In the mid-1980s, the organization had adopted a policy of dynamic support to specific technical measures for disease control to bring about a 'child survival revolution'. Support for 'basic services'—in which water supply and sanitation had prominently figured—and to organic processes of community change, was put on the back burner in favour of specific medical interventions—principally, mass immunization and oral rehydration therapy (ORT) for diarrhoeal dehydration in the under-fives. With the emphasis on the immediate saving of children's lives, the message to Unicef's engineers worldwide was that they had been relegated to the sidelines. Handpumps and toilets might be important for public health over the longer term, but in terms of dramatic and demonstrable reductions in young child deaths, they could not compete with medical technology. More could be done, faster and more cheaply, to save children's lives by needles, rehydration solutions, and basic medicines than by the preventive strategy of water points, toilets, hygiene, and health education.

Proportionately, Unicef's support to water and sanitation as a category of its assistance was dwindling. At the beginning of the Water Decade, Unicef globally spent more on water than on health. By the early 1990s, health consumed almost three times as much. Because the water programme in India was valued by the government and keenly supported by major donors, it managed to hold its own in spite of falling interest from Unicef policy-makers. Meanwhile, the momentum behind technical interventions for 'child survival' worldwide exponentially grew. The major preoccupation in every Unicef country and subsidiary office, including those in India, become focused on putting in places strategies and galvanizing resources for disease reduction in young children by means of mass immunization and ORT campaigns.

Thus the Unicef water and sanitation professionals in India were caught in the middle of conflicting currents. On the one hand, despite voices to the contrary, the Indian government was fixated on technological aspects—witness the very name of the Drinking Water *Technology* Mission, and its five sub-missions, all of which were technically focused—on fluoride, iron, brackishness, guinea worm, groundwater management. On the other hand, the international (or donor) drive was all for 'software' and 'process'. Meanwhile, Unicef itself was engaged in a major thrust based on technology, but this was exclusively medical technology for saving child lives as quickly and cheaply as possible, and embraced neither water or sanitation, nor had much time for education or behavioural change except around its key interventions.

However, there were cracks in the prevailing GOI and Unicef policy frameworks that could be opened. One of these was the attention to social mobilization and social marketing that accompanied the 'child survival revolution'. Even immunization, an intervention which is universally uniform, as far as the technology is concerned, required that mothers be persuaded of its virtues, and bring their children to the vaccination clinic. The case for using social marketing to carry across messages about water use and sanitation was put to the National Drinking Water Technology Mission (it later dropped the 'technology' from its title) and was persuasive. The Mission did have a limited health education role in its terms of reference—to carry out an 'awareness campaign' through the mass media. Although this was a limited approach, at least it was something.

As a preliminary to conducting this campaign, the Drinking Water Mission asked Unicef to commission a countrywide study on water use and hygiene in rural areas during 1988–9. Up to this time, all that the planners, policy-makers, and programme practitioners had to go on were their assumptions. They did not actually know anything other than anecdotally about their intended beneficiaries' views and behaviours. Without sound information, no sound social marketing strategy for the mass media campaign messages could be devised. This Knowledge, Attitudes, and Practices (KAP) study, conducted by the Indian Market Research Bureau, was the first serious effort to understand the parameters of water and sanitation-related behaviour in India and consider their influence on the incidence of disease. The study covered more than 7900 individuals in villages of 22 districts in eight states in different parts of the country, and sought answers to such fundamental questions as how people defined 'safe' and 'unsafe' water, 'cleanliness' and 'health', why they believed what they believed and did what they did in relation to water and sanitation. Many of the study's findings came as a shock, for example, the discovery that most people did not appreciate handpump water as 'safe' and preferred to drink the water from their traditional sources (see Chapter 3).

One of the study's important observations on hygiene concerned methods of water storage in the home (GOI 1990). In Gujarat, Madhya Pradesh, and Rajasthan, the water pot was often stored above ground level in a special niche in the wall or on a platform, to keep it out of reach of animals and small children, but elsewhere it was usually left on the floor. The pot was usually covered, but only in Gujarat and Rajasthan—thanks to the guinea worm eradication campaigns—was water filtered. In 68 per cent of households, water to drink was taken out of the pot by using a cup or vessel without a long handle, which meant that dirty fingers were dipped into the water. Contact between hands and water occurred frequently during collection, transportation, storage, and serving. This was one of the principal ways whereby germs were transmitted from person to person.

Another key finding relating to water was that, while 88–95 per cent of people across the states believed that 'bad' water caused disease, their view of why it was bad was more often associated with its appearance, taste, and whether it was good for cooking, than whether it might contain germs or whether it came from an open or protected source. They also thought that the illnesses it caused were fevers, coughs, and colds; only 10–18 per cent had any idea that it might be connected to diarrhoeal diseases and stomach disorders. As far as cleanliness was concerned, people had a strong sense of inner and outer purity, and made a connection between personal hygiene and disease; almost everyone believed it was important to wash hands

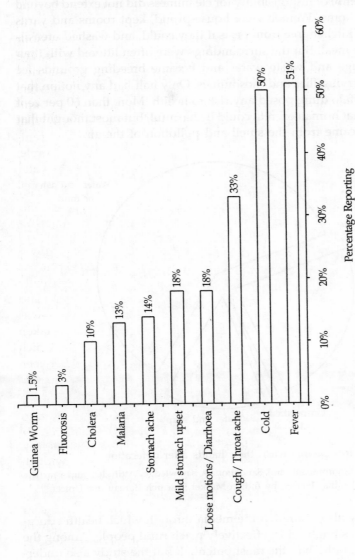

FIGURE 5.2: People's perceptions on health problems caused by bad drinking water

Source: Water, Environment, and Sanitation: A Knowledge, Attitudes, and Practices Study in Rural India, 1988–9, by Indian Market Research Bureau, for Unicef.

149

after defecation and before eating even if they didn't always do so. But their sense of responsibility for cleanliness did not extend beyond their own home. Women were house-proud, kept rooms and yards swept, the kitchen free from pests if they could, and washed utensils after every meal. But the surroundings were often littered with their own garbage and waste water and became breeding grounds for disease-carrying flies and mosquitoes. Only half had any notion that cow or buffalo dung posed any risk to health. More than 60 per cent believed that human excreta could be harmful, but most thought that the threat came from the smell and pollution of the air.

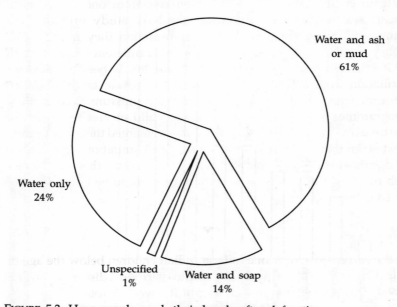

FIGURE 5.3: How people wash their hands after defecation

Source: Water, Environment, and Sanitation: A Knowledge, Attitudes, and Practices Study in Rural India, 1988–9, by Indian Market Research Bureau, for Unicef.

The study also examined the media through which health education messages might most effectively reach rural people. Among the mass media, radio had the most potential. But the study also underlined that interpersonal communications—through teachers, health workers, and others—were vitally important. It found that the only village-level functionary of the water supply and sanitation program-me—the handpump caretaker—was not carrying health messages to the community at all. It concluded that, since the improvement of health was its key objective and this could not be realized without tackling the poor understanding and range of unhygienic practices the study had uncovered, communication and motivational activity would now need to be given their proper status at every level, with adequate resources, specialized personnel, and time allocated to developing messages and putting them across.

For Unicef's water and sanitation professionals in India, trying in difficult circumstances to change their role from one of 'technical spearhead' to 'software advocate', the KAP study provided the ammunition and scientific basis of information they needed. The policy of supporting 'area-based' projects in different locations with NGOs and enthusiastic state partners had already been adopted for sanitation. It could now be expanded to include targeted health education messages. At the same time, the pressure to align their programmes with the key organizational 'child survival' priorities of the time could be met. From these dynamics emerged the programme's next major thrust—the coupling of water and sanitation with 'control of diarrhoeal disease'. Thus was born a programme that was neither fish nor fowl but tried like a mythical creature to be both—Control of Diarrhoeal Diseases (CDD)–Watsan.

In the early 1990s, over one million Indian children below the age of five died annually of dehydration due to diarrhoeal disease (UNICEF 1996). This represented one-quarter of the world's total of four million diarrhoeal deaths. Diarrhoeal cases also accounted for as many as 40 per cent of paediatric beds and more than one-third of paediatric out-patient visits in peak seasons of the year.

Since the mid-1980s, Unicef had been pushing hard for a transformation in case management for young children suffering bouts of diarrhoea as part of its 'child survival revolution'. The key to reduction in diarrhoeal deaths, Unicef believed, was to familiarize parents and health workers with ORT to offset the drastic loss of fluid which diarrhoea inflicted on a small body, threatening to drain away life itself. ORT's emphasis was on extra liquid and nutritional intake for sick children, and on home administration of ORS sachets which contained salt, minerals, and glucose to be mixed with water. The idea was to de-mystify diarrhoeal disease, and enable parents to handle the majority of cases effectively at home without recourse to expensive medicinal remedies, and without the weight loss and growth setback that were the frequent legacy of illness in the young child.

In India, as in an increasing number of countries, ORS sachets were being locally produced by pharmaceutical companies in increasing volume, and could be purchased or obtained from health centres. There had been heavy promotion of this remedy through the health infrastructure. Now came the idea that this could be done equally well through the Watsan programme. In fact, several Unicef water and sanitation programmes around the world were at this time becoming directly involved with diarrhoeal treatment so as to be seen as engaging in the organization's priority activities. Some felt they had no choice. When the Watsan programme in Pakistan came up for review in 1986, it found itself unfavourably compared to ORT as a means of controlling diarrhoeal disease, and fighting to remain in existence (Wurzel 1990). At the 1990 UN World Summit for Children, countries around the world had committed themselves to the target of treating 80 per cent of cases of diarrhoea in young children with ORT by 1995 and halving child deaths from diarrhoeal disease by 2000 (UNICEF 1994). India had by the early 1990s only managed to reach 37 per cent of cases. So the Unicef Watsan programme rode to the rescue.

The CDD–Watsan strategy was developed as a pilot programme and, following a baseline survey (UNICEF 1993), was launched in 15 districts in 1992–3. In each location there were three components: water supply and sanitation including safe disposal of human excreta, education on the causes and prevention of diarrhoeal disease, and management of childhood diarrhoeal cases. The health logic was impeccable. If water and sanitation programmes were supposed to reduce diarrhoeal diseases, why not tackle diarrhoeal case management as well, thereby fixing in people's minds a holistic view of prevention and cure? This required a 'convergence' of services in an 'integrated' programme framework—concepts then coming into development vogue and perceived as the last word in programme design. However, 'convergence' and 'integration' are difficult to achieve in practice since this is not the way services usually work, and the sharing of tasks and swapping of roles between workers from different disciplines and sectors has to be carefully organized or confusion rapidly reigns.

What happened was that the relative emphasis on the three components varied from location to location, depending on the energy and commitment of the various departments and personnel involved. Where the health department and medical officers were keen, the CDD effort came to the fore. In other locations, where departments of rural development or public health engineering led the way, water supply or sanitary marts were the main activity and CDD took a back seat (UNICEF 2000). Naturally, medical people do not go in for engineering installations, and engineers do not handle medicines. It is hard to see what the actual 'convergence' of services was from the perspective of those running them. However, the emphasis on CDD was welcomed in locations where making measurable reductions in diarrhoeal deaths among the under-fives with a medicinal response was easier to grasp and run with than spreading the more nebulous package of 'total sanitation'.

One of the CDD–Watsan districts was Erode in Tamil Nadu, previously scene of an area-based sanitation project supported by Unicef. Different sectoral departments—health, education, rural development, and the water board—were all involved (UNICEF 2001). Each department had its own focus. The health department attended to diarrhoeal disease control, the education department concerned itself with school sanitation, the rural development department looked after handpump installation and set up Sanitary Marts, the water board was concerned with well-drilling, monitoring water quality, and combating fluorosis, which was common in Erode district. Unicef provided the necessary financial and technical support for implementation, as well as assistance for training task forces at village, block, and district levels.

The strategy was 'training of trainers' and the passing down of skills through the district and block layers to the villages. The village task force consisted of 15–20 members, representing all the villages and hamlets around a given health sub-centre. Its members attended training sessions run by the district task force, and then set out to hold meetings in all the villages. *Kala Jathas* (street plays) were performed by trained troupes to help identify people's needs and priorities, prompt group discussion, advocate the construction of soak-pits and handpump platforms, and help set up ORT depots.

The CDD activity in Erode was highly effective. Between 1992 and 1997, nearly 500 health personnel from PHCs and government hospitals were trained in the prevention and treatment of diarrhoea. Over 6000 mothers attended one-day training programmes on the use of ORS and hygiene in the home. A similarly extensive programme of health, hygiene, and CDD training took place in the schools.

ORT corners were set up in all major health facilities, and for every health sub-centre, five ORT depots were established. The depot holders were village women provided with 10–15 sachets of ORS to give out to mothers of children suffering from diarrhoea, and trained to help administer the anti-dehydration mixture correctly. Each sachet cost a few rupees, a quarter of the price in a local shop, allowing the depot holder to replenish her stocks from the health sub-centre and make a tiny profit. Altogether, 635 mothers were set up as ORS depot holders. This meant that basic treatment to prevent diarrhoeal dehydration was close to the doorstep of every patient. Over the course of six years, health centre data showed that cases of diarrhoea had declined. Data from two PHCs in Dharapuram taluk showed that the total number had declined from 1263 in 1992 to 635 in 1997. Five deaths had taken place in Dharapuram in 1992, but none between 1995–7 (UNICEF 2001).

The pilot CDD–Watsan schemes, with some stops and starts, ran through most of the 1990s. Many borrowed from the methodology of the SWACH project. Village contact drives focusing on diarrhoeal disease were the backbone of the project in Allahabad district in Uttar Pradesh, for example (UNICEF 1996). IEC through songs, sketches, wall-paintings, posters, and group discussions were central to the social marketing of diarrhoeal disease messages. However, in many locations convergence and integration remained elusive. One aspect was given prominence and others ignored.

In some, interest was limited to the availability of Unicef funds to pay for drilling of additional boreholes and India Mark III handpumps. To boost the preventive impact of safe water supplies, the density of handpump installation in CDD–Watsan areas had been increased from the standard norm of one pump within 1.6 kilometres to one pump within half a kilometre. This made the programme attractive to state PHEDs, because, in the knowledge that they had no resources to cover such an expansion, the financial burden fell on Unicef. This inhibited the potential for replication beyond the experimentation period, when Unicef funds for drilling and Mark III handpumps would no longer be available. The luke-warm attitude of many states to the programme once hardware subsidy was eliminated was undoubtedly one of the reasons that the programme fizzled out in many of the original CDD–Watsan states.

A number of reviews were conducted by donors to the different CDD–Watsan locations. In Orissa and West Bengal, the Department for International Development, the British donor, found evidence of a drop in the incidence of diarrhoea, and a reduction in the need for hospital referrals because of the availability of ORS in village depots (UNICEF 1998). Everywhere depots were found to have a steady clientele, and apprehensions that they might not be viable proved invalid. Hygiene promotion had also proved successful through schools and anganwadis. In Medinipur, sanitation was forging ahead. CDD–Watsan had been an umbrella under which various useful initiatives had been launched, but it had not conjoined disease control and public health engineering in a formula configuration guaranteed to enable water supply, sanitation, and diarrhoeal disease control to move successfully forward. Everyone was doing their own thing.

Whatever had been achieved had not come about as a result of the linking of CDD to water and sanitation. Although many mini-successes were achieved through the CDD–Watsan approach, it did not demonstrate particular strengths as an integrated health, water, and sanitation strategy. Demand for it did not develop in other districts or states. When the 15 pilot schemes came to an end, CDD and ORT were no longer integrated or adopted as the vanguard of water and sanitation programmes. 'Convergence' of services had not managed to prove itself more effective than standard sectoral approaches—at least not yet. It had not proved possible to develop a strategy comparable to the one for guinea worm eradication with an onslaught on diarrhoeal disease as its leading edge. Questions about this remain unanswered. Was it because of lack of official commitment and incentive to engage as strongly with the software side as with the hardware of handpumps and medical technology? Was there no community demand? Was the strategy wrong? Or is the spectrum of diarrhoeal infections and their multiple routes of transmission more complex to address?

Significant reduction of diarrhoeal disease in infants and small children, and in other illnesses associated with poor environmental conditions, still proves elusive. There has been a reduction over the past decade, from over one million diarrhoeal deaths in the under-fives to around 400,000. But it is not clear that water and sanitation are the critical factors, however important they may be as a pre-condition. The two National Family Health Surveys conducted during the 1990s (1992–3 and 1998–9) confirmed that little impact had been made on childhood diarrhoea merely by the provision of safe drinking water sources (UNICEF 2002). The provision of home toilets, however, did bring about a 20 per cent reduction in mortality among the under-twos. These findings bear out the continuing need to put across health education messages: protecting drinking water all the way from source, to home, to storage pot, to consumption; hand-washing after defecation and before cooking, eating, or handling food. Unicef is now focusing on school hygiene and sanitation as the most likely way to bring about a transformation in behaviour and reductions in disease.

On the basis of scientific understanding, it has become clear over the years that the continued promotion of safe rural drinking water supplies cannot be justified exclusively on their capacity to combat diarrhoeal disease and death. Nonetheless, this continues to happen without the necessary attention to hygiene education needed to secure the potential child health benefits. This affects the credibility of water and sanitation programmes, which would be well advised to reconsider the emphasis on disease control—too often presented as their sole legitimization. The best that can be said of repeated analyses of associations between the provision of water—less so of hygienic sanitation—and reductions in diarrhoea is that they are open to interpretation. The variables in disease causation are far greater than any such mono-focal analysis can capture. Clearly, safe water supplies are fundamental to good health, and to the practice of hygienic behaviour. But it is a serious mistake, perpetuated by the international health community over many years, to justify water and sanitation inputs on health impacts alone. When the promised reductions in disease do not materialize, the instinct of some, in Unicef at least, is to argue for the withdrawal of support to rural water supplies. This is to downgrade their importance in enabling people to live comfortable, dignified, and minimally adequate lives on grounds which are, essentially, specious. The obsession with health also ignores the fact that access to a safe and dependable water supply is essential to life in all sorts of other ways than merely its direct impact on morbidity loads.

The missing link between water, dignity, life, and health can only be provided by education and hygienic behaviour. In turn, the success of health education and calls for behaviour change depend on provision of water and sanitary facilities. Without these, how can hands and bodies be washed, food or utensils be cleaned, excreta be flushed, clothes be laundered, or minimum standards of personal dignity be maintained? The absence of water also means, for many families living at the margins of existence, a serious depletion of livelihood chances, compromising household food security prospects and affecting child health in other ways. Many studies have endeavoured to prove the links between water supplies and better health, and have come up with disappointing results. But the reality is that no healthy or productive life can be lived without water, whatever the studies say.

REFERENCES

Black, Maggie (1990), *From Handpumps to Health*, UNICEF, New York.
Government of India (1990), *People, Water and Sanitation: What they know, believe and do in rural India*, National Drinking Water Mission, GOI; based on information contained in the report *Water, Environment and Sanitation: A Knowledge, Attitudes and Practices Study in Rural India, 1988–9*, IMRB and UNICEF.
UNICEF (2003), *Banishing the Guinea Worm: The Story of Eradication*, India Report.
————— (2002), 'Child's Environment Programme (2003–7)', Programme Plan of Operations.
————— (2001), *Guinea Worm Eradicated from India: The End of a Scourge*, WHO and UNICEF; also containing quote from 'Report on Surgical Extraction of Guinea Worm—the AIIMS team'.

————— (2001), 'Field Note on Control of Diarrhoeal Disease', Child's Environment Programme, No. 12, 2001 series, UNICEF India Country Office (draft).

————— (2000), *Learning from Experience: Evaluation of UNICEF's Water and Sanitation Programme in India*, 1966–98, Evoluation Office, UNICEF, New York.

————— (1998), *Control of Diarrhoeal Disease, CDD–Watsan, Fourth Progress Report*, March 1998.

————— (1996), 'Case study on Community-based Water Quality and Surveillance Monitoring under the CDD–Watsan Project', Allahabad, Uttar Pradesh, UNICEF Field Office, Lucknow.

————— (1996), *Control of Diarrhoeal Diseases: Water and Sanitation Project (CDD–Watsan)*, First Progress Report, Government of UK and UNICEF, April to December.

————— (1994), *The Progress of Nations*, UNICEF, New York.

————— (1989), 'Apex Committee Working Group on Health Education and Community Participation Report (1986)', quoted in Gordon Tamm, Rural Water Supply and Sanitation in India, Sectoral Achievements and Constraints, UNICEF Consultant Report.

Wurzel, Peter (1990), *Maximising and Sustaining Health through Water Supplies and Sanitation—the Pakistan Experience*, Water Quality Bulletin, Vol. 15, No. 1, January; quoted in Maggie Black, *Children First: The Story of UNICEF*, UNICEF and Oxford, 1996.

6

Managing a Community Service

Evidence suggests that in the distant past the management of local water structures and installations in India was a community affair, as was the management of land, rivers, forests, and the whole natural resource base (Agarwal and Narain 1997). The complications of managing water for farming in the ferocious monsoon climate meant that village societies all over the sub-continent enjoyed a high degree of hydraulic literacy many centuries before they could read or write. Elaborate systems of water harvesting, conservation, and allocation—as varied as they were intricate—existed in every setting, from the tempestuously flood-prone to the arid semi-desert.

Less is known about how they were run from the social, as opposed to the technical, point of view, and less is known about the management of water for drinking than for irrigation. But the thread of local construction, maintenance, and responsibility over the systems on which people were dependent can be traced, through poems, stories, songs, and myths handed down through the generations, as well as from historical sources. So important was water that it was regarded as sacred. Rivers such as the Ganga and the Narmada continue to be revered as holy. Traditionally, the digging of a tank, a well, a canal, or a reservoir was regarded as one of the most meritorious acts a devout Hindu could perform for the community during his or her lifetime.

The innate sense of community responsibility for water sources infused official thinking about water supply programmes in the early years after Independence. When the government's rural drinking water supply programme first got into its stride in the early 1970s, an assumption was made that the gram panchayats, the traditional custodians of community water installations, would take the new handpump–boreholes into 'ownership' and become responsible for maintenance and repair (see Chapter 3). This never happened. So, when maintenance issues began to be seriously addressed from the mid-1970s onwards, apart from developing a handpump that would be less susceptible to breakdown, responsibilities for maintenance and repair were instead entrusted to teams of government engineers. The only tasks assigned to the community were to manage the handpump area and report breakdowns to the authorities. Still, assumptions remained that the handpump–borehole would be seen as a community facility, and that it would occupy pride of place in terms of drinking water source. Clearly this mechanical device which produced clean water from a protected underground source would be regarded as superior to the old-fashioned dug well, exposed to the elements and to detritus casually entering from all kinds of direction. This idea turned out also to be over-simplistic. The assumptions on which it was based cut across existing 'water wisdom', underestimated consumer priorities, and ignored attitudes about who 'owns' and has responsibilities over facilities provided by the government.

Instead of seeing 'their' handpump as a community installation, two out of three users, according to the KAP survey undertaken in 1988–9—believed it to be the property of the government (GOI 1990). This belief was particularly strong in the northern and central parts of the country. In Andhra Pradesh and Gujarat, a significant proportion (30 per cent and 41 per cent) thought the pump belonged to the gram panchayat. In some states—West Bengal, Uttar Pradesh, Gujarat, and Tamil Nadu, and particularly in Manipur—a noticeable proportion of people did think the new installations were 'public' facilities, and did expect to bear some responsibility for maintenance and repair.

Contrary to expectations, people also said they were prepared to pay a certain amount regularly towards maintenance of the pump. In West Bengal, 89 per cent of respondees were willing to pay, 81 per cent in Manipur, and in Uttar Pradesh, Andhra Pradesh, and Gujarat, 70–80 per cent. This was a very surprising finding. The government service providers had taken it for granted that the people would refuse to pay anything at all, even where water was scarce. This assumption was based partly on the idea that they could not afford to do so. It also stemmed from the perception that communities would be resistant because they had been promised a free water source, free in the same way that traditional sources were 'free'.

In fact, people's expectations were not so crude. The findings of the study as a whole underlined that the relationship between rural communities and their water sources, and what they expected from the government or did not, was a great deal more complex than had been thought. Dug wells were still first on the consumer choice list as far as drinking was concerned, partly because the water tasted better and often because they were located nearer. In spite of the involvement by the government in water supply schemes, both for irrigation and for drinking water, a sense of autonomy was retained by communities over traditional sources, including their management. They deepened dug wells and de-silted tanks as needed, and where suction pumps had been in use for generations, kept them in repair. Meanwhile, there was some ambivalence about India Mark II handpump water, except in places where or at times of the year when there was no alternative supply. However much welcomed, as all water supply services were bound to be anywhere in India, those provided by handpump–boreholes were often viewed slightly askance. Well-drilling and handpump technology had been absorbed effectively by the government, the industry, and 'the market', but the services they provided had not been absorbed into the fabric of rural society. There were many factors involved, but one was unquestionably the lack of communication and mutual understanding between users and providers.

A survey conducted in 1986 on the performance of India Mark II handpumps had found that the reporting system on breakdown was 'uniformly poor', irrespective of the maintenance system (Samanta et al. 1986). Around 20 per cent of handpumps were out of action. In half the cases of current failure, the problem had yet to be reported. When asked why they had not bothered to report the failure to the authorities, the usual reply was that no one knew whom to go to— even where caretakers were in place. Eventually some local user reported the breakdown after several days, a week, or even longer.

This implied that the community's dependency on the handpump supply, and the corresponding sense of urgency about its failure to function, was not of a high order. Worse, the authorities' sense of its importance to the community was even more dilatory. In 72 per cent of breakdown cases found by the surveyors, no one had visited the pump site. In the other cases, someone had visited but had lacked tools or spare parts to do anything about it. Weeks might pass until the repair team showed up. This did little to instil confidence among community members about the dependability of the handpump source. It was still seen as just one of various alternatives, and so long as the other alternatives held up, they did not feel sufficiently exercised about a breakdown to do something urgent about it.

By 1988, 1.4 million India Mark II handpumps had been installed countrywide (UNICEF 1998). The problems surrounding their maintenance had not been adequately resolved, and as numbers were rapidly increasing, the centralized system—three-tier, two-tier, one-tier, or whatever adaptation was employed—would become more overburdened as every year passed. On cost grounds alone, this expansion was not indefinitely sustainable. The maintenance system cost an average of Rs 350 a pump, and in spite of this expenditure, remained inefficient. Many pumps remained out of action for unnecessarily long periods due to failure to undertake small repairs and a lack of rudimentary spare parts. Unicef, therefore, became convinced that a radical approach to operation and maintenance (O&M) should be pioneered. A system for community-based handpump management was devised, and in 1988 it began to be piloted in eight varied locations in seven states: Madhya Pradesh, Andhra Pradesh, Maharashtra, Rajasthan, Uttar Pradesh, West Bengal, and Assam. As usual, the idea was that when the approach had been tested and fine-tuned using Unicef resources, a streamlined version could be disseminated across the country at government expense and used as a model for community-based maintenance of water supply services generally.

Apart from the need to devolve some of the costs and to increase efficiency, the article of faith underlining Unicef's new maintenance approach was that the supply-driven nature of the programme had eclipsed the participation of community members in service delivery, and that this failure lay at the heart of the disinterest they felt about it. The theory was that if they became responsible for the maintenance of handpumps and were obliged to contribute towards it, they would be motivated to make sure that repairs were efficiently executed. In addition, since some of them would have been trained to do the work involved, they would also be 'empowered' to carry out repairs and maintenance. Instead of being seen as 'beneficiaries', users would become informed and discerning consumers and managers of services.

The most daring feature of the approach was the role to be accorded to women. Not only were they to be involved as users and managers, but they were to be trained as village mechanics. Women would be given employment skills and opportunities in preserves traditionally confined to men. This had already been pioneered by SWACH. Now it was to be elevated from a position of eccentric deviation in programme design in tribal areas of Rajasthan, to an integral feature of the new countrywide model for community management of handpumped water supplies. What were the indications of possible success?

The SWACH project managers in Rajasthan had been committed to finding locally devised and managed solutions for development problems. But although the existing 'one-tier' system of handpump maintenance adopted as the state policy was supposed to work in this way, the SWACH managers had found it highly unsatisfactory in Banswara and Dungarpur districts. When a handpump broke down, villagers were supposed to report it to their *panchayat samiti*[1] member, who would in turn inform the sarpanch at the next samiti meeting. He was then supposed to contact the local village mechanic or mistri to undertake the repair. In practice, the panchayat samitis had not undertaken these responsibilities effectively, the mistris were often uncooperative, and loopholes for enabling fraudulent claims to be made on the government budget for repairs which had never taken place were frequently exploited (Mehta 1993).

---

[1] The panchayat samiti is the panchayat council body which operates at a level above the village panchayat, and to which all villages send one representative.

SWACH believed that these problems might be overcome by training female mechanics because women as the prime users of
handpumps would be sincerely committed to maintaining them and
responsive to the needs of other users. An additional advantage
would be that if they could function effectively as mechanics, Rajasthani
women would break the mould of expectations concerning their
capabilities, in their own minds and those of their menfolk. For the
first time, they would be able to earn money in their own right in a
recognized 'job'. In 1989, SWACH organized the training of 24 women
in Banswara and Dungarpur districts as handpump mechanics, and
shortly afterwards, a further 24 women were trained in Udaipur
district.

Altogether, the 48 women were expected to look after 480 handpumps, one mechanic for every 10 pumps. Using bicycles and tool-
kits with which they were provided, they were to work in threes,
looking after 30 pumps in adjoining villages. This was because they
had to travel considerable distances between far-flung communities
carrying a heavy tool-kit on foot or bicycle. Outings of this kind on
their own would be unheard of for women, and seen as exposing
them to impropriety or worse.

One of the most difficult aspects of the scheme was to persuade
women to undertake the training, to allay their own fears as well as
overcome family resistance. An evaluation of the programme undertaken in 1993 found that SWACH had been instrumental in the
process of persuasion. A high proportion of the women recruited—
over three-quarters—had previously been animators for the guinea
worm eradication programme and had, therefore, already gained
confidence about undertaking 'working' functions outside the home.

The evaluation set out to discover whether the idea of training
women mechanics had proved viable and effective, what their motivation and background was, the attitudes towards them of their families and communities, how they performed in comparison to male
mechanics, and whether they were likely to stay in their jobs, even if
they had to work without a government stipend. The overall tenor of
the evaluation was positive. After three years, 40 out of 48 women
handpump mechanics were still in their posts—a more than respectable retention rate. In spite of feeling that they received too little pay—
Rs 11 per month per handpump—87 per cent wanted to continue,

which was a very important finding. They had not been demotivated by their low pay to consider abandoning their jobs. Therefore, the other benefits they had gained from the programme—higher status, new knowledge, the opportunity to go out and do something reward-ing—compensated for the poor remuneration.

The picture painted by the report revealed a great deal about the women, and served to overcome many prejudices about whether they were suited to this new role. All came from marginal farming families, and three-quarters were from scheduled castes or tribes. They tended to be younger than male mechanics and came from families with larger landholdings and higher income. Around two-thirds were educated above 8th grade, and it was clear that the higher the educational standard, the better the performance. Since they came from large families, when they were away being trained or out on their handpump 'calls', their domestic tasks were redistributed among other household members, so this had not proved to be a problem.

Some had encountered initial opposition from families or relatives, but once they had been working for a while, resistance tended to dissolve. An extra source of steady income, however small (an av-erage of just over Rs 2000 per year), was not a matter of indifference for families whose annual household income was typically less than Rs 18,500 (US$ 300) and, in many cases, a great deal less. They looked after fewer pumps than men, and so their earnings were lower. But their performance at their task was comparable to those of the men. Most people looked up to them. They had gained in confidence and self-respect. However, many of the male mechanics did resent them (UNICEF 1993). They did not want women to muscle in on work they would otherwise have had. And they did not receive free bicycles and tool-kits. The cost of a tool-kit was very high, amounting to several months' earnings. This was felt to be grossly unfair.

There was also a lack of support from the PHED, whose assistance was needed when repairs—for example, to replace old and rusted leaking pipes—were beyond the village mechanics' competence. And there were problems with spare parts, pumps running dry in the hot season, and indifferent support from the panchayati samitis. Women or no women mechanics, community maintenance of handpumps was not an undiluted success. Many people still preferred to drink and cook with water from their traditional dug wells, which remained the backbone of water sources on which people felt they could depend. However, the more general problems associated with community maintenance and the sustainability of handpump supplies were eclipsed by the positive findings about the effectiveness of the women mechanics.

So enthused was Unicef by their performance in Rajasthan that the training of women as village mechanics became a lodestar of the community-based handpump maintenance schemes introduced in other states from the late 1980s. The 'women factor' was assigned the role of the latest magic ingredient which would make community maintenance work. However, it was an uphill struggle to persuade government officials of its virtues, even if the women themselves were keen. Take Betul, a hilly district to the south east of the Satpura mountains in Madhya Pradesh. Here, there was a strong bias among officials in favour of training young unemployed men. By 1993, only a handful of women had been trained as mechanics, and even the women themselves admitted that social attitudes, the distances they had to travel to the villages on their beat, and caste and class divisions made their jobs and acceptability difficult. However, their families supported them, as did the local PRIs, and communities came round to the idea once they had proved themselves to be tough enough and capable. Income was, as ever, an important motivation for the women and their families and, equally, could be a reason for disapproval or resentment from others.

Betul was not the only project district where training and setting up of women handpump mechanics was slow to get off the ground. When the 1993 evaluation of the projects took place, there were many communities with no mechanics, male or female. In Rangareddy, in Andhra Pradesh, where the three-tier maintenance system had long been established, a number of active women handpump caretakers were found, but very few women mechanics. Those that existed were in what were known as 'demonstration areas' which had received generous amounts of development funds under various programmes because they lived in the constituencies of prominent political figures. 'Project' areas had received almost nothing, and very few of their handpumps were in working order. None of the women interviewed in 'demonstration areas' seemed to have repaired or maintained more than one or two handpumps since their training two years earlier. However, the women were enthusiastic, could remember how to dismantle the pump, and others were eager to join the scheme. Again, they were keen to take on any job or earning opportunity, however modest.

The significant problems with community maintenance of water supplies—and there were clearly quite a few—did not relate to women's involvement. Where women had been involved—in Rajasthan, Andhra Pradesh, and West Bengal—this was one of the few positive features. Women in some of India's villages had taken on technical responsibilities and acquitted themselves well. However small a breakthrough in rural women's emancipation, it was definitely something. It was the one bright spot in what was otherwise proving a dismal record as far as community maintenance of government-supplied rural water services was concerned.

After the International Water Decade of the 1980s, there was considerable push from bilateral and international donors to persuade recipient countries to abandon supply-driven approaches and reduce the government's own participation in delivering and running services. Instead, the devolution of service management to communities, user groups, and private bodies, including NGOs, was recommended. This was promoted on grounds of making services more responsive to people's needs, better run, and more cost-efficient. In 1992, an International Conference on Water and the Environment in Dublin came up with a set of policy principles to guide the debate on water resources management and conservation at the first Earth Summit, held at Rio de Janeiro later that year. The Dublin principles together with those articulated in Agenda 21, the key policy document to emerge from the Earth Summit, have remained the overarching principles put forward by the international community for water policies ever since.

The concern which began to dominate international thinking at this time was that of environmental constraint. From the recognition that fresh water was a finite resource under increasing pressure, and would have to be conserved in order to serve the needs of humankind in future generations, flowed a number of other policy consequences. The classic way to constrict the use of any resource is to raise its price. Since the industrial revolution of the late nineteenth century in Europe and elsewhere, water and sanitation services had been heavily subsidized from the public purse in virtually every country and state, partly in the interests of public health, partly in recognition that the lowest-income members of society could not manage without water and could not afford more than a minimum fee. By the late twentieth century, with water resources in need of protection, the subsidies over-spend and the lack of cost-effectiveness, efficiency or elementary appreciation of conservation needs led to a call for drastic change. In future, subsidies should be reduced and the costs of water, especially the costs of engineering works to harness, store, or transport water, should be passed on to consumers.

Even poorer citizens would not be immune. The fact that many people in the poorer parts of African and Asian towns and cities were currently paying quite a significant proportion of their income to water vendors was cited to suggest that there was no need to be as beneficent as water supply administrators had been in the past. Cost-recovery became the new theme of water supply programmes, whether at the macro investment level for municipalities or in the more modest world of rural handpump schemes. Establishing mechanisms for community participation in service management was one way of setting up a user fees collection system. It was also a way of ensuring that consumer voices would be given weight in service choice, which might make cost-sharing more acceptable.

Therefore, one of the most important principles to emerge from Dublin and Rio concerned the need for participatory processes and structures. Agenda 21, the action document agreed to at the Earth Summit, recommended that delegation of water resources management be at 'the lowest appropriate level'. This should be done in accordance with 'national legislation, including decentralization of government services to local authorities, private enterprises, and communities'. In the future, users should be involved in the choice of technology and sites, the implementation of water and sanitation schemes, and manage the service to the degree of which they were capable. Impeccably democratic and accountable though this sounds, and sincerely intended to enable communities to exercise consumer

---

Box 6.1

The 'Rio Principles'

- Ensure the integrated management and development of water resources.
- Assess water quality, supply and demand.
- Protect water resource quality and aquatic ecosystems.
- Improve drinking water supply and sanitation.
- Ensure sustainable water supply and use for cities.
- Manage water resources for sustainable food production and development.
- Assess the impact of climate change on water resources.

Note: These principles formed the basis of chapter 18, Freshwater Resources of Agenda 21, the Earth Summit's key output, Rio de Janiero, 1992.

Box 6.2

The 'Dublin Principles'

- Fresh water is a finite and vulnerable resource, essential to sustain life, development, and the environment.
- Water development and management should be based on a participatory approach, involving users, planners, and policy-makers at all levels.
- Women play a central part in the provision, management, and safeguarding of water.
- Water has an economic value in all its competing uses and should be recognized as an economic good.

Note: These principles were articulated at the International Conference on Water and the Environment, held at Dublin in January 1992.

---

control, introducing participatory management systems into societies characterized by social and economic inequality is fraught with complication.

Participation in services starts with siting. In order for new hand-pumps to be readily accessible, in India they are often installed alongside roads on waste or public land. But rarely does discussion occur about the desirability of one site or another, or its convenience for the majority of its users—the women. Even where it does, 'community' decisions in traditional rural areas may not be reached by methods considered in modern parlance as 'participatory' or 'democratic'. The voice of authority, and those of the better-known and better-off landowners in the area, usually hold sway. This may have its advantages. Powerful and more educated householders are more likely to take action promptly if a handpump breaks down, and they may also have the means to subsidize repairs from their own pockets. Alternatively, they may exercise enough influence at the block or district headquarters to ensure that repairs are done properly. On the other hand, they may also act to defend their own interests, and oversee the management of 'community' services in such a way that the poorer majority of the community are effectively excluded.

Anyone with experience of community-based handpump maintenance, in India or anywhere else, soon loses the illusion that, however desirable an improvement on centralized O&M, there is anything simple about introducing participatory service management into the typical rural setting. The same is true of panchayati raj reforms intended to lead to greater democratization of decision-making in local affairs. There has rarely been a tradition of listening to people of inferior status or to women as far as community matters are concerned, or of giving them any kind of equal say. This is part of an entrenched pattern of social organization and cannot be dissolved in days or even years. Often, the active involvement of a powerful and public-spirited local leader is essential, otherwise his counterpart—the non-public-spirited leader—will take over and run local bodies as he sees fit. The creation of user organizations strong enough to survive when the original impetus to come into being is removed, is the surest guarantee of success. But there is never any guarantee of enduring success. Each community has to find its way, and not all will manage to do so. No foolproof formula for community management of water or any other basic service can ever be devised. This is a problem that technocrats with their guaranteed solutions to technical problems, and the bureaucrats used to holding sway, find difficult to accept. They find it even more difficult to adjust in mid-course in order to make a system work.

The community-based system Unicef promoted, apart from women mechanics and handpump caretakers, consisted of the following elements: village-level Watsan Committees (with 50 per cent women members); maintenance funds, to which users would contribute a monthly levy; the use of NGOs for mobilization, information, and education; and the introduction of VLOM handpumps. The state government bodies responsible for rural drinking water were to provide back-up, training, supervision, and carry out major repairs. The package was very ambitious given that the post-Water Decade thrust of the national rural water supply programme in India had not yet altered course to match international changes of policy direction. It was still very much hardware and target driven. Numbers of physical installations, rapid drilling, and handpump coverage were the order of the day. Thus, state and district government institutions were not really committed to community-based maintenance with all the difficulties it entailed. State PHEDs were apprehensive that this type of exercise would slow down the pace of coverage, and they would fail to meet politically-driven targets. They also scorned the idea that communities could develop the necessary technical competence, nor did they welcome the loss of control that community-based management was bound to entail.

A key decision, whose full implications were not foreseen, was to install India Mark III handpumps in the project areas, either as new facilities or to replace existing Mark IIs. The reason was simple. In order for village mechanics to be able to handle maintenance and elementary repairs, the handpump would have to be fully VLOM. Some states resisted this on grounds of extra costs. In Andhra Pradesh, the Panchayati Raj Engineering Department was not prepared to undertake the conversion programme until community maintenance had proved successful—a Catch 22 situation. In Rajasthan, the authorities preferred to upgrade the Mark IIs by changing the cylinder and washers, only using the Mark IIIs in new installations. In Pune, Maharashtra, the water table was so low in the summer months that extra pipes had to be installed for the Mark III. Their weight, combined with poor quality installation, led to frequent breakages at the pumphead, and the pumps became unpopular or unusable.

Thus, even technology conversion—the most straightforward part of the programme—did not go easily. But the worst problem was another classic Catch 22 situation. New Mark IIIs or converted Mark II pumps broke down so rarely in the initial phase of their lives that there was nothing to do for newly-trained and enthusiastic village mechanics or committees. In fact, as far as the government bodies were concerned, the projects were largely regarded as opportunities for testing out the Mark III technology. Their interest in the social and institutional side of community maintenance was lukewarm at best.

It gradually became clear that the only way that genuinely partici-patory systems could be developed was by relying on NGOs to shoulder not just the educational and awareness-building activity, but the development of the whole social infrastructure—as the Ramakrishna Mission had done in West Bengal. Medinipur, in the early 1990s, was one of the project areas for community-based handpump maintenance, as well as for intensive sanitation programmes (see Chapter 4). At this time, the new direct-action Tara pumps were being installed in West Bengal. The sanitation project in Medinipur was, therefore, deliberately interconnected with provi-sion of water supplies. Once 40 families in a village had built toilets, they were entitled to the installation of a handpump. Under the guidance of Ramakrishna Mission field workers and youth associa-tions, the communities made regular contributions to maintenance and opened a savings account. The passbooks were held by women caretakers, and repairs were carried out by local mechanics, also female, and trained by the Mission. So efficient was cost recovery that each village, if they kept their payments up, would be able to afford a new pump when the initial one wore out (the Tara's lifetime is around 10 years) (Black 1996).

Few NGOs, however, enjoyed the same reputation, length of ex-perience, and local influence that the Ramakrishna Mission exer-cised. They not only had to have grassroots empathy and understanding to work closely with communities, but they also had to work with block officials—the development officer, as well as teachers and health and anganwadi workers—and manage to exer-cise some leverage higher up. One example of a Unicef partner NGO was Water Education and Social Action (WESA) in charge of coor-dinating the project in Betul, Madhya Pradesh. WESA had many motivated young men and women staff who spoke the local tribal dialects. Adopting the idea of Village Contact Drives, they visited their project villages several times, staying with the people, holding meetings in halls and anganwadi centres, conducting door-to-door campaigns, showing slides, putting up posters, holding musical entertainments, and so on. They created good relationships with the people, set up user groups and Watsan Committees, established bank accounts, trained the members, and did everything by the book. But all of this activity did not make the scheme work. Many of the handpumps were dry or yielded insufficient water for reasons beyond the NGO's or the community's control. The PHED was uncooperative. Their mechanics failed to show up and no spare parts were available. The villagers lost confidence in government officials, and the NGO was itself highly frustrated (UNICEF 1993).

In Maharashtra, a longstanding and highly professional NGO existed, under whose umbrella 70 smaller local NGOs were affiliated. This was AFARM, a body that had helped bring hard rock drilling into India, and was also the key in the development of the precursor handpumps to the India Mark II. AFARM was also experienced in community mobilization, awareness-building, and action. However, its supervision or seal of approval did not necessarily guarantee any better results than those in Betul. One of the local NGOs, an AFARM member called Mahila Vikas Prakalp, put forward a proposal to organize community maintenance for 100 handpumps. The Zilla Parishad obliged them to take up all the handpumps in the block. Having agreed to provide Rs 200 per pump plus spares, the Parishad reneged and supplied only Rs 100. The NGO went ahead anyway, managing to achieve 95 per cent maintenance success. They were then subjected to press and political criticism, told to stop work, and had all their payments frozen. Meanwhile another NGO working in the same locality, with no experience of water supplies or handpumps, had the patronage of the authorities and faced no problems. When the project was assessed against its targets, it was clear that the only area in which there had been reasonable progress was in the installation of India Mark III handpumps.

Altogether, between 1988 and 1996, Unicef promoted 24 community-based handpump maintenance projects in a variety of settings. There were undoubtedly some achievements. Fewer handpumps remained out of order. Village mechanics became responsible for 80 per cent of repairs undertaken, and women became recognized as effective mechanical operatives. Although these outcomes were satisfactory, they masked many problems about back-up, spare parts, accountability, and whether the systems put in place with such effort and at such considerable Unicef expense would survive for any length of time (UNICEF 1998).

Whether intentionally or simply by default, the idea that the lower echelons of the government bureaucracy could introduce village-level services in such a way that they would properly belong to or be run by the community had been more or less abandoned. The international donors, including Unicef, pushed ever harder for social inputs and transformations, cost recovery, community maintenance, sanitation, and health education, while the PHEDs pursued coverage targets ever more relentlessly. Thus, the role of NGOs in creating the necessary institutions and receptiveness at local level became critical. In effect, the government handed over its responsibilities. But there was also official ambivalence towards the growing status of NGOs. Even though many officers at subsidiary levels worked closely and harmoniously with them, at higher levels there were reservations. There was annoyance with international donors at their inclination to side-step the bureaucracy in favour of NGOs. This was in spite of the fact that everyone accepted that NGOs were better at reaching communities and interacting productively with them. Typically, though not invariably, they also had more integrity of developmental purpose where the seriously disadvantaged were concerned.

One ground-breaking initiative was the Swajal project in Uttar Pradesh. Here the World Bank provided the lion's share (84 per cent) of the US$ 71 million required to provide water and environmental sanitation to 1000 villages in 19 districts, most of which were poor and many in hilly tribal tracts (World Bank 2000). These costs were extraordinarily high—US$ 71,000 per village—compared to the PHED norm. The strategy depended on selecting NGOs, which tendered for the contracts, to work with villages in planning and constructing their schemes. There were choices in technology: gravity feed, rainwater harvesting, wells, handpump–boreholes, depending on whether communities were in the hills, foothills or the plains. For the better-off, piped schemes were offered with individual household connections for those who could afford to pay a premium. Whatever type of system or system combination they selected, communities had to provide 10 per cent of the capital costs of installation, and thereafter assume 100 per cent of maintenance costs. There was to be no contracting by the state or district water bureaucracy. Instead, communities were to procure building materials themselves, choose any necessary construction firm from those they trusted, organize community labour, and supervise the whole operation. Responsibility for the services was allocated to PRIs.

The Swajal project was definitely visionary within the Indian context, and the forcefulness with which it was promoted by key donors had an impact on policy attitudes in the government. However, in addition to its high costs, it also had a flaw. It was a time-bound experiment, and it presumed too much about the effectiveness of its model for devolving operations and all maintenance costs onto communities before it had been properly tested. Swajal's stated objectives in Uttar Pradesh were to provide sustainability for rural water supply services, improve rural incomes and women's status, and confer health and sanitary living on the communities involved. Over and above these, the Swajal project was also to 'test alternatives to the current supply-driven service delivery mechanism'. This was meant for the whole country, not just for Uttar Pradesh.

The project's implementation phase did not begin until 1996, and the construction of water systems throughout all the 1000 villages would not be complete before 2002. Except in a preliminary way, Swajal's impact on health or its capacity to fulfil its various decentralization objectives would not be ready for full evaluation until or beyond that time. The project was outside the regular machinery of government, and once it ended there was no guarantee of follow-up. Thus there were questions about the replicability and sustainability of the strategy, from its inception.

By this time the move for 'sector reform' in Indian water supply and sanitation had reached a crescendo. In the late 1990s, the central government, with encouragement from international donors including the World Bank, decided to put into place a policy for sector reform based on the new ideas exemplified in Swajal. There would be a 'paradigm shift' from supply-driven to demand-responsive approaches; local communities would be 'empowered' to participate in the implementation and operation of drinking water supply schemes; women would be closely involved in decision-making; the government would become a facilitator instead of a provider (GOI 2000). This was the standard donor recipe for using market mechanisms to encourage conservation of resources and user participation in service management. There were many officials and politicians in India who recognized the financial and democratic virtues of decentralizing services. But resistance was also to be expected. Vested political and bureaucratic interests would be affected.

The way the central government decided to implement 'sector reform' in water supply indicated that some of the dilemmas were recognized. But the new thinking was predisposed towards the way the Swajal project had attempted to resolve them. An assumption was made that this was a proven template for community-managed and market-based water services delivery. But had Swajal really cracked the puzzle of how to proceed towards decentralization of rural water services, let alone equity in delivering them to the least well-served? In all reality, it was too optimistic to draw any such conclusion.

The preamble to 'sector reform' for water and sanitation services came with the two amendments (the 73rd and 74th) to the Indian Constitution passed in 1992. The changes these made in the status, composition, and powers of local councils, for rural and urban areas respectively, were far-reaching. The rationale behind their introduction was that the old PRIs governing local affairs had been withering away over time and, in many states, had become weak and ineffective. They no longer held regular elections, spent their time wastefully, and were unrepresentative of the weaker sections: women, scheduled castes, and tribes. Part of the problem was that, in modern India, their functions were ill-defined, and they had insufficient access to financial resources from taxes and central government allocations, whose dispensation was in the hands of politicians and the bureaucracy.

The two constitutional amendments were expected to make good these deficiencies and institute a profound democratization and devolution of power to communities and neighbourhoods, through gram sabha and other elected intermediary panchayati bodies up to the zilla parishad level. Among the 29 policy and programme areas in which these bodies were, in future, to have decision-making powers were water resources management, drinking water supplies, health and sanitation facilities, educational institutions, and programmes for poverty alleviation. One-third of seats on the panchayati bodies was reserved for women, and there would have to be members, on both the councils and in office-holding positions, of scheduled tribes and scheduled castes in proportion to their representation in the population at large. Stipulations were laid down about the conduct and regularity of elections (every five years), to all offices (The Constitution 1992).

The transfer of water and sanitation responsibilities to these lower-level institutions corresponded to another of the central orthodoxies of the post-Rio international water and sanitation agenda—that government bodies should be facilitators for, not providers of, services. Eminently sensible and inoffensive though this sounds on paper, in its implementation it requires divesting the existing loci of power and expertise of many of their current responsibilities. It therefore also implies downsizing government structures and pay-rolls—certainly part of the agenda promoted internationally by the World Bank with all the leverage at its command.

With their jobs and patronage at risk, PHEDs and other bodies in charge of development programmes and budgets were not likely to lie down quietly and disappear. And they might well have a case. The gram sabhas and panchayats were embryonic in terms of management skills and not yet equipped with the expertise to make techni-cally-sound or effective decisions. Installations under their ownership might fall into disrepair even more quickly than they had in the past. Meanwhile, politicians who had promised free water supplies and other amenities as part of their election platform would not lightly give up the possibility of manipulating votes by such promises, however poorly fulfilled and mistrusted they had become. Such a profound upheaval in the distribution of power and financial re-sources to lower levels could not be engineered overnight by decree.

To begin with the transfer of responsibility over water supply services was nominal rather than real. (There were important excep-tions, for example in Kerala and West Bengal, where local democracy and PRIs had been promoted since the 1970s.) Gram panchayats elsewhere tended simply to implement programmes handed down to them by state and central governments, and were still under the sway of the grant-making control exercised by government bodies pulling the strings (Nigam et al. 1998). In practice, for the first few years, most state governments still acted as providers of basic services in rural areas and showed little inclination to give up their decision-making authority, let alone the patronage with which it went hand-in-hand and to whose fruits many officials had become habituated.

In order to work, decentralized management of drinking water supplies required a radical change of attitudes. Since state water boards and PHEDs would not lightly allow their responsibilities to be fragmented and devolved, they would have to be actively persuaded, even coerced, into abandoning much of their authority and powers. They would have to be able to see some positive advantage in ceasing to run the show, and instead demonstrating to others, less well-educated and socially respected, how to do so with their support. The professionals—engineers in particular—would have to be cajoled into passing on technical expertise instead of jealously guarding it. Orientation would be needed at all tiers for officials and panchayati officers to understand their new roles and enter into them with enthusiasm instead of reluctance. The new structures would have to be adapted to local political, economic, and ecological circumstances, and take into account social and cultural mores, including the situation of women. Instead of being imposed by those 'in charge', services were to be designed and run in a spirit of dialogue and interaction with ordinary people, taking their views into account. This would be a tall order anywhere in the world, not just in India.

In 1999, the sector reform policy was announced, to be carried out by the National Drinking Water Mission.[2] Initially 67 pilot districts were designated as 'sector reform' districts, in which the new decentralization of service management to communities was to be tried (GOI 2000). By involving communities in the planning, implementation, maintenance, and repair of existing schemes, and their replacement or expansion where needed, the idea was to establish a sense of ownership and responsibility, and ensure effective management

---

[2] In 1993, the National Drinking Water Mission was renamed the Rajiv Gandhi National Drinking Water Mission in honour of India's assassinated prime minister, who had set up the original National Technology Missions in 1986.

and service sustainability. The formulation for funding was that 90 per cent of the capital costs would be provided by the central government, while 10 per cent of the capital costs and all the maintenance costs were to be shouldered by the communities themselves—the same formula as Swajal. In each of the pilot districts, a District Water and Sanitation Mission—a sub-body of the National Mission—was created to oversee the implementation of sector reform projects. The total costs of this exercise were quite substantial—Rs 18 billion (US$ 390 million). This would cover the extension of participatory management of water supplies to around 70 million rural people in 10 per cent of the total rural area of the country. It was expected to generate Rs 1.8 billion (US$ 39 million) from the communities. Once tried and further improved on the basis of experience, the approach would be extended throughout the rest of the country.

In each 'sector reform' village, there would be an awareness campaign and capacity building, carried out by an NGO, to generate demand for a project. A village water and sanitation committee would be set up to plan and implement construction and organize O&M. This committee would be part of the gram panchayat and under its control. Echoing the provisions of the 73rd and 74th Constitutional Amendments, places on the committee would be reserved for women and members of scheduled castes and tribes. The time allowed for project execution was 36 months, permitting a good period for thorough sensitization, training, planning, and implementation. Once the new scheme had been commissioned, the community would own it legally, know how to manage and repair it, and be fully aware that the government would not step in and repair it if it broke down. They would also appreciate that they would need to collect sufficient funds for maintenance themselves.

Unfortunately, 'sector reform' was seen in many minds as primarily an exercise to introduce cost-sharing, however much its rhetoric included software components—information, education, community planning, and the rest. And without those components, reform would be synthetic and understood by communities as merely a means to make them pay for services which, well-run or poorly-run, used to be provided free of charge. Given that they had often shown themselves ambivalent towards these services, preferring to depend on traditional water sources where they could, and neglectful of maintenance and repair, the idea that they should contribute towards the costs must initially have seemed odd. They had been given an installation that they had not requested, and were then asked to pay for it. A lot depended on how desperately handpump water was needed, and this naturally varied from location to location and user-group to user-group, even within one district or block. It also depended on whether a community or a group within a community had the resources to make the necessary contribution.

Then there was the other side to the picture—the top-down system of introducing policies and programmes which bureaucrats usually prefer. Many state and district entities had not yet undergone the necessary conversion in hearts and minds to make decentralization work. They took the new template and imposed it, much as they had the old one. Instead of contracting companies to drill holes and build installations, they contracted NGOs, some of which were effectively companies in another guise, to undertake the projects. They in turn worked with officials and gram panchayats. The gram panchayats were often dominated by local landowners and other men of influence, and were, especially to begin with, far from models of equitable and democratic power-sharing. The important members would rarely choose to neglect their own self-interest in favour of the landless, low-caste, and other disadvantaged groups. Where women were elected to key positions because this was a part of the new panchayati rules, they were often the puppets of their husbands—at least in the early years of the reforms.

In some cases, the effect of decentralizing service management was positive; in others it was anti-democratic. Where little in the traditional countryside is run on truly participatory lines, but is still steeped in the hierarchies of caste and status, weakening the role of the state authorities could undermine the cause of social equality. Their programmes had previously intervened in such a way as to compensate for gross inequity—for example, by focusing on 'problem villages', or acting on behalf of women, scheduled castes and tribes, or 'below poverty line' groups. Where government policy laid down specific targets in these contexts, there would be some sort of conformity, however imperfect. When the government no longer had the responsibility to intervene on their behalf, weaker sections might be unable to defend their interests. If a system was poorly installed or technically inadequate, for example, they might not have sufficient consumer clout to demand redress against local authorities or contractors in their pay.

The problem was exacerbated by the fact that piped water schemes were of much more interest both to the influential people within the communities and to the authorities. Rural landowners with aspirations looked down upon the lowly handpump. It was seen as a facility only for the use of the landless or poorer sub-groups. Moreover, among government officials, patronage dynamics favoured larger schemes with more construction, especially where central government funds were involved. In some better-endowed states—Tamil Nadu for example—virtually every village had a piped water supply scheme and handpump water was now looked upon merely as a secondary source (GOI 2000). Since the community was only required to provide 10 per cent of the capital costs for any scheme, larger and more prosperous villages could manage 10 per cent of a piped water scheme, whereas seriously disadvantaged communities would have difficulty in affording 10 per cent of any type of scheme. Thus, the likelihood was that the Drinking Water Mission resources for 'sector reform' would be spent on better-off communities, which would then become even more privileged than the poorer ones.

The results of 'sector reform' could, therefore, be described as patchy at best. Six years down the line, with the Swajal project about to be wound up and its institutional innovations about to confront the need to survive on their own, enthusiasm for its 'sector reform' model has become diluted. Take-up in 'sector reform' districts has not been what it should, and adjustments have been made to encourage more villages to come forward with project proposals. A one-off grant of Rs 50,000 (US$ 1000) will now be given to each project community as a revolving fund for community maintenance start-up; and the proportion of installation costs to be paid by scheduled caste and scheduled tribe villages (that is, villages where more than 50 per cent of the population is in these groups), has been reduced to 5 per cent.

From the perspective of the central government, there can be no going back on the decentralization process. Whatever the shortcomings of its implementation, it has become accepted that sustainability of both water resources and water supply systems requires community-based management, local contribution, and ownership. Although the general picture is less encouraging than had been hoped for, the principles of the new policy remain valid. Correspondingly, in late 2002 the government decided to adopt a proactive role and push for reforms with greater vigour. On 25 December 2002, under the banner of *Swajaldara*, the then prime minister, Atal Behari Vajpayee, formally announced that the reforms would be accelerated, and would no longer be confined to the original 67 reform districts, but extended throughout the country.

Meanwhile, in pockets of some blocks and some districts, good things have happened. This gives hope that, where there is genuine local commitment from users and change-agents alike, community management of water services is not an impossible dream. Let us return to Malikpur village in Tonk, Rajasthan (see Chapter 4). Tonk was not a district originally designated for sector reform, but it has made considerable progress towards community management of services. Since 1999, Unicef has been supporting a water supply and sanitation programme which promotes local responsibility for water supply systems through the gram panchayats. In Tonk, most gram panchayats cover four or five villages, or 1000 families altogether. A suitable NGO is assigned to a group of panchayats, and it acts as the conduit between villages and the programme. Its job is to train village mechanics and health animators, establish user groups for each handpump, organize elections of water and sanitation committees, liase with block officials, and generally nurse the programme through its paces. When the institutional basis for community management is secure, the NGO withdraws, and the committee can call upon local officials of the PHED when needed.

The NGO working with the people of Malikpur on water supply and sanitation is a well-established local organization called Centre for Development Communications and Studies (CDECS), which is based in Jaipur about 120 kilometres away. Suraja Karan Chaudhary, sarpanch of Malikpur, recounts how CDECS first contacted them two years ago, and adds: 'We have really benefited from that contact.' There are four handpumps, each with its user group, in the village. 'Each group collects funds from the user families, and they use this money for repairing the handpumps. This they do with the help of the local mechanics who live here and have their own businesses. It has not taken more than 24 hours to carry out any repair so far.' There are three handpump mechanics in the village, one man and two women. When a pump breaks down, people tend to call upon the person nearest to them, male or female. The most common repair is to change the cup washer on an India Mark II or III. This takes around two hours, and they charge Rs 40 (less than US$ 1). If the repair is beyond their competence, they bring in the local PHED.

Everyone seems satisfied with the system. The account books stating the contributions and expenditures of all the user groups are there for inspection. The names of all the families are registered, their contributions listed, and no sanitary detail of their lives—toilets, soak-pits, smokeless stoves, drinking-water care—is omitted. The 61 families in one user group have collected Rs 436 and so far expended Rs 312 on maintenance. All speak highly of the support they have received from CDECS officers, with whom they have regular meetings alongside government development staff at the block headquarters. In due course, CDECS will bow out, leaving the system, hopefully, intact. The strength of this approach—a Unicef hallmark—is that it is fully integrated with the government structure. It is not some project set apart, as in the case of Swajal or SWACH, which ends at a given time when all its staff disperse to other jobs and its offices close down.

Here at Malikpur and in other communities in Tonk, not only are the institutional mechanisms being given thorough attention over the short and longer term, but there is a strong commitment to the programme from the government and a high-class performance from the NGO. The socio-economic and environmental contexts, too, are important for this achievement. The enthusiasm of the users, animators, mechanics, and committee members is palpable. Rajasthan, under many diverse influences, has been a crucible of community activity for water supplies and water resources management. Not least because the people in Rajasthan have a very high degree of motivation to maintain and manage every single source of water available to them.

Community management of water supply—handpump or other— is by no means looking so good everywhere in the country. Results are still disappointing in many settings. The mistake is to assume that this is a type of intervention which, if one can only fine-tune the design template, will simply roll out and 'go to scale' as with a handpump or toilet. It is possible to develop a template for building community institutions, complete with principles, practices, and time-frames, with which no one could take issue. But the most perfectly designed social programme is only as good as those who implement it. Unlike technical items or consumer gadgets, it cannot be made human-error proof. Human capacities and failings are the key ingredients of whether organizational systems work or do not. Good technical components are preconditions of success, but they cannot guarantee anything on their own.

There are many Indian NGOs and government officials who, if allowed to do so, implement programmes well and wisely, and genuinely involve people in the communities, taking account of their views. Theirs should be the decisive voices in how to implement community management of services in any given setting. Hopefully, in the future, that is the direction 'sector reform' for community management of water and environmental sanitation will increasingly take, as more people understand what a 'paradigm shift' to 'demand-responsive' services genuinely entails.

REFERENCES

Agarwal, Anil and Sunita Narain (eds) (1997), 'The Rise and Fall of Water Harvesting' in *Dying Wisdom: Rise, Fall and Potential of India's Traditional Water Harvesting Systems*, chapter 3, p. 281, CSE, New Delhi.

Black, Maggie (1996), 'Water, Environment, Sanitation: The Changing Agenda' in *Children First: The Story of UNICEF*, chapter 4, OUP and UNICEF.

Government of India (2000), 'Community Participation in Rural Water Supply', Indian Initiative, Rajiv Gandhi National Drinking Water Mission, Ministry of Rural Development, GOI.

——— (1990), *People, Water and Sanitation: What they know, believe and do in rural India*, National Drinking Water Mission, GOI and UNICEF, New Delhi.

Mehta, B. C. (1993), *Involvement of Rural Women in Water Management: Scheme of Women Handpump Mechanics: An Evaluation*, Society for Development Research and Action/UNICEF, Jaipur.

Nigam, Ashok, Biksham Gujja, Jayanta Bandyopadhyay, and Rupert Talbot, *Freshwater for India's Children and Nature*, UNICEF and WWF, New Delhi.

Samanta, B. B., Dipak Roy, and T. N. Dutta (1986), *Survey on the Performance of India Mark II Deepwell Handpumps*, Operations Research Group/ UNICEF.

The Indian Constitution (1992), The Seventy-Third and Seventy-Fourth Amendment Acts.

UNICEF (1998), *UNICEF-Watsan time-line*, prepared for Evaluation team, UNICEF Water and Environmental Sanitation Section, New Delhi.

———— (1993), *Evaluation of Community Handpump Maintenance Projects in Seven States with Special Focus on Women's Involvement*, and also including chapters on Betul and Maharashtra, a DCI report, UNICEF, New Delhi.

World Bank (2000), Welcome to the Swajal Project.

7

Threats to Water—and
Everything Else

Over the decades since 1972, when India launched its rural water supply programme, the volume of investments absorbed by the programme rapidly expanded, and the coverage of facilities rapidly increased. According to the government's statistics, by the end of the century the results had brought India within a hair's breadth of 'Water For All'. By 1994, 95 per cent of India's rural people were said to be 'covered' by the water supply programme, according to the Ministry of Rural Development. This did not mean a tap in every home or even a standpipe at every corner but, all the same, it was a major achievement. Almost everyone appeared to have access to a reliable source of safe water providing 40 litres per person per day, at a distance of no more than 1.6 kilometres.

This figure of 95 per cent coverage of the population was reached by a crude system of reckoning. Over the years, the numbers of 'problem' villages and habitations (hamlets) without a safe water source had been identified by government surveys. When they subsequently received an installation under the Rural Water Supply (RWS) programme—deepwell handpump–borehole, piped scheme, gravity-flow, or shallow tubewell—they were designated as 'covered'. However, to state that 95 per cent of India's villages had access to a safe water source implied that only 5 per cent of India's villages had a drinking water problem. Could this possibly be true?

In December 1999, N. C. Saxena, previously a secretary in the Ministry of Rural Development and then secretary of the Planning

Commission, presented a paper at a State Water Ministers' Workshop in Cochin. He took the opportunity to point out some strange features of these coverage figures. Surveys of 'problem' villages and habitations had taken place at regular intervals since 1972. Even though a large number of villages had been covered by the programme between each of the surveys, the number of 'problem' villages and habitations always went up rather than down (Saxena 1999). For example, after the work undertaken between 1985 and 1994, out of 161,722 'problem' villages in 1984, only 70 should have been left without a supply. But when surveyed again in 1994, there were 140,975 problem villages. So instead of 95 per cent coverage, as had already been announced, there was suddenly considerably less than this. As Saxena put it: 'In our mathematics, 200,000 problem villages minus 200,000 problem villages is still 200,000 problem villages' (Agarwal 2001).

At this time, the Ninth Plan (1997–2002) was underway, and the Rajiv Gandhi National Drinking Water Mission was reporting good progress towards what would soon be full coverage in 'not covered' (NC) or 'partially covered' (PC) habitations. (NC meant that there was no source, whereas PC meant that there was only an unreliable source, often dry in the summer months, or that some other norm was unfulfilled.) By the end of 1999, according to Drinking Water Mission figures, the outstanding proportion of 'uncovered' would again be 5 per cent of the rural population. But no survey of facilities in any part of the country supported the view that only 5 per cent of villages were without drinking water. On the contrary, about half of the habitations faced acute hardship or severe water quality problems, according to Saxena's estimates. 'Why this extraordinary discrepancy between government records and reality?' was the question he posed (Saxena 1999).

A number of reasons could be put forward for the see-sawing numbers of 'problem' villages down the years. One was that the norms for defining 'problem' villages had been refined at various times to provide more water per head of population; to ensure small settlements—habitations—had a source; or to increase water output in livestock areas (70 litres per head per day instead of 40 litres) (Black 1990). A second reason was that surveying, testing, and assessing water supply outputs had improved, and was undertaken with greater regularity. Another reason was that there was such political impetus behind the programme that it was in everyone's interests to maintain its momentum, by endlessly altering its 'coverage' goalposts if need be. But there were two central reasons for what amounted to a continuing drinking water crisis, despite the huge Plan expenditures absorbed by the programme over three decades—a poor record of maintaining the facilities, and long-term unsustainability.

Saxena quoted a Planning Evaluation Organization survey in 29 districts conducted during 1996–7. Although the number of people who had access to public drinking water sources had risen from 69 per cent to 81 per cent in 10 years, the surveyors found a litany of problems. Two-thirds of the districts reported frequent water scarcity; 40 per cent of the villages surveyed had shortages in the summer months; 30 per cent reported that their new service was undependable and were forced to fall back on their old water sources. In the case of piped water supplies there were frequent power breakdowns, leakages, and contamination. In the case of handpumps, poor construction in nearly half of the cases and frequent pump or borehole failure was the problem. The involvement of local people was conspicuous by its absence. In only three villages out of 87 had a Watsan committee been formed, and in no village could any trained village mechanic be identified. Problems of poor quality installation—boreholes, pumps, platforms, drainage systems—and shortcomings in maintenance were to blame for the frequency and duration of breakdowns. There were shortages of staff and resources too in certain states. In one district, Bijnore in Uttar Pradesh, there were only 11 mechanics to look after 4000 pumps.

But there was another, ultimately even more fundamental, issue. The resource itself was under increasing stress. Inadequate facilities and weak maintenance infrastructures were not the only villains. The best management systems in the world could not easily compensate for declines of 15–20 metres in the water table over a few years. This trend was accelerating, while programmes for groundwater recharge were only tardily being introduced and languished far behind what was needed. India's agricultural and municipal water supply policies were sucking the country dry. Handpumps and piped supplies were increasingly collapsing in the summer months because there was no water. A 1998 study in eight districts of Madhya Pradesh found that nearly 30 per cent of handpump villages and 88 per cent of piped water supply villages were no longer fully covered (FC) as designated, but partially covered. Up to 10 per cent of handpumps became defunct in any year, which was not being taken into account in the coverage statistics. In other states, such as Andhra Pradesh, the accelerating trend for 'fully covered' villages to become 'partially covered' villages was also a major cause of concern. All over the country, pumps and taps were running dry but, according to the Drinking Water Mission, since pumps and taps existed, water supply coverage was nearly universal. There was clearly something wrong, both in analysis and prescription.

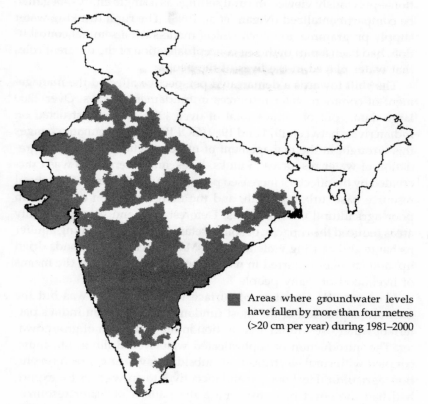

Areas where groundwater levels have fallen by more than four metres (>20 cm per year) during 1981–2000

Map 7.1: Groundwater Depletion Areas of India
Source: Central Ground Water Board, February 2004.

The decline in India's water table had begun in the 1960s and 1970s, when food scarcities and the rapidly growing population had burdened India with drought and undeclared famine. It was at this time that many of the old drinking water sources—shallow wells and suction pumps—had begun to dry up, and the high-speed drilling rig with its instant access to lower water depths in hard rock areas was heralded as the instant purveyor of water supply salvation. The need to expand food production, and the advent of the miracle seeds of the Green Revolution, had encouraged government policies towards increased irrigation for agriculture. For the first time, the use of water for drinking and household supply, and the use of water for irrigation—previously viewed in rural settings as a single entity—began to be compartmentalized (Nigam et al. 1998). The rural drinking water supply programme, to which Unicef made such a seminal contribution, had been a part of this re-conceptualization of the different roles that water played in the lives of the people.

The shift towards a demarcated perspective affected the management of common water resources in fundamental ways. Over 2200 large dams, out of India's total of over 4000, were constructed on Indian rivers between 1971 and 1989 (GOI 1994). The dramatic changes this wrought in the management of upper catchments led to degradation of water resources in tanks, lakes, and rivers. This was exacerbated by the effects of increased population, pressures on the natural resource base, urban growth, and the marginalization of people in poor agricultural environments. Deforestation and erosion in hilly areas reduced the capacity of soil to retain rainfall, preventing aquifer recharge and causing violent floods. Many streams and ponds dried up, and changes occurred in natural ecosystems affecting the means of livelihood of many people.

However, the alteration of surface water resources was but the outward expression of the most fundamental change in India's pattern of water use—rapid acceleration in the extraction of groundwater. The introduction of sophisticated waterwell drilling into India, coupled with rural electrification, subsidies favouring 'green revolution' agricultural technology, and incentives to grow crops for export, had had the effect of transforming the pattern of water resources exploitation throughout the country over the last two decades. The policy increased agricultural output, export revenue, and national food security—but at a heavy price. The modern drilling technology brought in for drinking water purposes had actually taken off industrially because of the demand for large-diameter boreholes and piped systems, mainly for irrigation.

Irrigation is a much thirstier affair than drinking and washing. Bolstered by a favourable regime of land and water rights, weak regulations, and low tariffs, groundwater extraction for irrigation went ahead unchecked. In parts of Gujarat, fossil water continues to be tapped (GOI 1994), and in some hard rock areas aquifers are now completely exhausted. India's water supply is, quite simply, running out. Centuries of sustained recharge would be required to replenish the deepest of the drained and empty fractures.

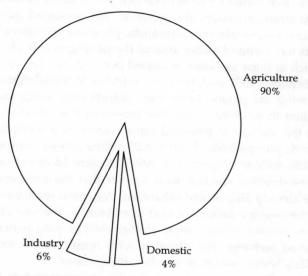

FIGURE 7.1: Users of water

Source: Earth Trends (2001) World Resource Institute.

Even in drought-prone states such as Gujarat and Maharashtra, farmers are encouraged to grow water-guzzling cash crops such as sugar cane. Similarly, rice—a crop which uses three times as much water per acre of cultivation as sorghum and 2.4 times as much as wheat—is widely grown in Karnataka and Tamil Nadu, two states which do battle every year over water releases from the Kaveri river.

Subsidies especially favour the better-off farmers, those with higher acreages; but such is the parlous state of the Indian farming economy for those on its lower rungs that, if the subsidies were removed, millions of small farmers would go to the wall. Even in the relatively modest rural areas of drought-prone states like Rajasthan, farmers grow water-intensive crops like cotton and oilseeds—so integral have cash crops become to their contemporary farming life. As the water table descends and their sources run dry, their own complicity in the disaster ahead seems irredeemable. The few states that have made tentative efforts to regulate groundwater extraction by legally limiting borehole drilling (Maharashtra and Karnataka) have found it very difficult to enforce the laws. In states where electricity prices have been raised (Madhya Pradesh and Andhra Pradesh), there has been strenuous political opposition. The conjoined interests of politicians, large-scale agriculturalists, planners, engineers, and bureaucrats have wound a knot around excessive groundwater exploitation which is none too easy to untie.

Between 1951 and 1990, the number of installations of all kinds drawing on groundwater grew significantly—dug wells from 3.9 million to 9.5 million, shallow tubewells from 3000 to 4.75 million; but the number of powered pumps grew by a spectacular amount: electric pumps from 21,000 to 8.22 million, diesel pumps from 65,700 to 4.36 million (Nigam et al. 1998). In some blocks of some districts, over-extraction reached such a point that the groundwater supply was already 85 per cent exhausted. A natural resource was, in parts of the country, being doomed to extinction. Without effective legislative control, conservation, changes in cropping patterns, and watershed recharge, the situation with regard to the exploitation of India's water resources was becoming chaotic.

Less than 5 per cent of groundwater extraction is for domestic water supplies—although 80–90 per cent of domestic supplies in rural areas and 50 per cent in urban areas are dependent on groundwater. However, in many poor, vulnerable, and drought-affected districts, it is difficult to make a hard and fast distinction between the importance of water for domestic use, and its importance for food production. Much of village India has always relied on irrigation of one kind or another, the rains being seasonal and unreliable. When the harvest fails, a family can survive on stored food for a period, but once this is exhausted and there is no alternative source to be had, there are few parts of the country in which a water source which is purely for domestic drinking purposes can be regarded as sufficient. In the absence of food, emigration, at least temporarily, to the city, a relief camp, or somewhere to assure survival becomes inevitable— at least for those who can manage the journey. Every year in India, in obscure corners of the landscape, people die whose resources failed and who could simply go no further.

When the National Water Policy of 1987 was issued, it had emphasized the need to regulate groundwater extraction, and to promote the recharging of groundwater aquifers. However, although a 'model bill' for this purpose was issued in 1990 and subsequently revised, little effort was made by even the worst-affected states to introduce appropriate legislation (Panda 2001). In Maharashtra—the only state to do so—a Ground Water Regulating Act was introduced in 1993, prohibiting the drilling of a well within 500 metres of a public water source. But despite good intentions, the Act was operationally confined to protecting public drinking water supplies, and had no application to conserving groundwater reservoirs as a whole (Chadha and Sharma 2001). In 1996, following a directive of the Supreme Court, a Central Ground Water Authority (CGWA) was set up to give greater force to the preservation of the national freshwater resource. The creation of this new body by no less an authority than the Apex Court was welcomed. But it was only a preliminary step.

In 2002, a revised National Water Policy was issued. Again, the policy stated that drinking water should be the first charge on any available water source. It repeated that the exploitation of groundwater should be regulated 'so as not to exceed the recharging possibilities'. There were also references to water allocation 'with due regard to equity and social justice'; the need for watershed management; conjunctive use of surface and groundwater; the need for participatory approaches involving women and Water Users' Associations; and a number of other progressive policy statements (GOI 2002). But without a radical change of laws, policies, attitudes, and investment patterns towards water resources management and conservation, it was hard to see how these principles could be put into practice. Unless the dynamics which lowered water tables were checked, the drinking water crisis was likely to continue, with more 'problem villages' emerging on a daily basis. The mounting pressures on groundwater and surface water needed desperately to be contained—to sustain both drinking water supplies and livelihoods, generally. This is a growing challenge with which many hard-hit states are trying to grapple.

The problem is not simply one of water quantity. The precipitous fall of the water table has also led to serious declines in water quality. The famous 'safeness' of groundwater for drinking has become distinctly less safe and the water less palatable, in an extra twist to the crisis.

Water quality issues first began to emerge in the late 1980s, and originally their focus was on bacteriologically suspect supplies. The great virtue of groundwater as the drinking water of choice was that it did not flow across, or stand about on, the surface of the landscape collecting germs and other detritus. It was naturally filtered through soil and rock, and came straight out of the earth clean and pure. As time went on, realization grew that it could become contaminated in between pumphead and lip. This problem became the target of educational messages about protecting the household drinking water supply as part of the Unicef-assisted environmental sanitation programme. But this was not the whole bacteriological problem.

Unicef's handpump survey of 1986 reported that as many as 44 per cent of pump sites were waterlogged because of poor location and drainage (UNICEF 2000). Where dirty water stood around the top of the borehole, contaminants from such water could pass through it. This could happen for various reasons. The borehole had not been properly sealed, the casing pipe was insufficient, or no handpump platform had been built. After the waterwell drilling industry took off in the 1980s and most of the increasing numbers of public boreholes were drilled by private contractors, it was difficult to ensure that specifications for installation of handpumps were fully met and standards of borehole construction maintained (see Chapter 2). The team of independent evaluators who, between 1998–9, examined Unicef's 30 years of assistance to water and sanitation in India concluded that most handpump–boreholes were not completed to quality standards—

usually because contractors took short-cuts. To add a sanitary seal to the borehole, for example, required additional time and expense. The contractor had to wait for the cement seal to dry, while his rig stood idle. This type of short-cut, therefore, meant that the source itself, not just the water once out of the ground, ran the risk of contamination (UNICEF 2000).

When samples from tubewells in West Bengal were tested in 1992, 29 per cent were found to be bacteriologically contaminated and unsafe to drink (AIIHPH 1993). Samples from handpump–boreholes in Allahabad showed lower rates of contamination (around 10 per cent) but even this was unacceptable (UNICEF 1996). Sanitation-related diseases, especially diarrhoeal infections, were continuing to take a heavy toll on young children's lives. In 1992 the estimate was one million. It was necessary, therefore, to address the causes from as many different directions as possible.

Under the umbrella of some of its existing community-based water and sanitation projects, Unicef decided to embark on experiments in community-based water quality testing. Till now, Unicef had been only minimally involved in this part of the national RWS programme, but as the problem became increasingly serious and it was clear that testing was not being adequately carried out, it decided to get involved. Unicef's approach was to look at the problem from the perspective of the community. The Rajiv Gandhi National Drinking Water Mission had decided to set up water quality testing laboratories in all the country's districts in 1990. But bringing samples to the laboratory from distant villages was bound to present a problem. Thus, Unicef began to focus on available techniques for testing water quality in the community itself. A number of simple devices were beginning to come on the market or were available from abroad, and different state programmes explored different possibilities.

In West Bengal, the All India Institute of Hygiene and Public Health (AIIHPH) and Unicef-Calcutta[1] decided to develop a community-based project model for water quality monitoring. The Institute was entrusted with running the pilot. A handy test kit, cheap and user-friendly, was developed for application in the field. Predictably, it was tried out in the redoubtable testing ground for all Unicef-funded

---

[1] At this stage, Calcutta had not yet been renamed Kolkata.

community-based water and sanitation action in West Bengal—
Medinipur. Of Medinipur's 54 blocks, 21 were covered by the Integrated Child Development Scheme, with its pre-school centres—
anganwadis—and anganwadi workers. These workers were designated as the core of the water quality monitoring initiative. The local
village panchayats were also important allies, as were members of
local clubs and block development workers, already tuned in to safe
water and intensive sanitation by the Ramakrishna Mission Lokasiksha
Parishad (UNICEF 1996).

The field kit for water testing developed by the Institute had to
be portable, low-cost, and easy for a villager to use. More than 10
existing designs were evaluated, but the Institute thought they could
do better. Through a long process of testing in the field and the
laboratory, the design was finalized in 1993. To make the kit lighter
and easier to carry about, the incubator needed to enable faecal
bacteria to mature was kept separate. It was also adapted for use with
a kerosene lamp instead of requiring an electricity supply, which was
not available in many villages. Manuals in Bengali were prepared for
the training of the anganwadi workers, who picked up the testing
procedure easily within the course of a one-day training. This entire
technology was put in the public domain, with the chemical composition of the reagents and a list of places where they could be obtained.

In June 1992, the pilot project began in 11 villages of Daspur I
block. Nine anganwadi workers monitored the water quality of all
drinking water sources—tanks, tubewells, ponds—for a period of a
year. The accuracy and reliability of the tests were checked by the
Institute, and found to be 98 per cent accurate (AIIHPH 1992). If an
anganwadi worker found any source to contain bacteria after testing,
she reported it to the block office, and measures to chlorinate and
disinfect the source were then undertaken. In 1993, following the
success of the pilot, the project was extended to 19 other rural blocks
in Medinipur. Within the next two years, 99 anganwadi centres had
received testing kits, 195 villages were covered, and the drinking
water sources of a population of 160,000 people were monitored and
kept free of contamination.

However, bacteriological problems were only one part of the increasing concern with water quality. In 1986, when the National
Drinking Water Technology Mission was first established, some of

the several sub-missions were concerned with chemical contamination of water supplies. Brackishness or salinity, iron, fluoride, and arsenic affected the safety or taste of groundwater sources in different parts of the country. Salinity was a problem in a number of coastal areas—Kachchh, in Gujarat, and Orissa and West Bengal, for example—where over-extraction of groundwater had caused serious salt-water intrusion. High concentrations of iron, found in parts of Bihar, Kerala, Madhya Pradesh, Tamil Nadu, Orissa, and Uttar Pradesh, was not harmful but it made the water taste unpleasant and ruined clothes laundered in it. Where people objected to how a new water supply tasted, they voted with their mouths for the old sources, not appreciating that these might be contaminated by bacteria or chemical ingredients with no taste, smell, or distinguishing colour.

After over 20 years of the national RWS programme, village India still did not know what a 'safe' water supply consisted of, and how an unsafe supply might bring contagion or undermine health. As the threat from contaminants in an over-stressed and increasingly populated environment grew, it became more urgent to put across that message. By the mid-1990s, it was estimated that 44 million people in India were affected by water quality problems of one kind or another.

Early in the 1990s, West Bengal became the source of real fear about water quality. Several districts began to find arsenic in their groundwater. The first cases of arsenic poisoning from tubewell water had been identified by the School of Tropical Medicine in Kolkata as early as in 1982. This was the first time that the natural contamination of groundwater by arsenic from a geological rather than industrial cause had been encountered. Scientists still differ as to the precise set of geochemical circumstances which triggers the absorption of naturally occurring arsenic into groundwater. But the lowering of the water table by excessive water extraction was widely regarded as reinforcing—if not triggering—the problem. By the mid-1990s, arsenic in groundwater was beginning to affect wide swathes of the West Bengal countryside around the mouths of the Ganga River. In 1997, WHO declared that arsenic poisoning was a major public health problem in the region, with eight districts and 13 million people at risk. Across the border in Bangladesh, delays in researching and addressing the phenomenon caused a furore in national and international circles which has not yet fully died down.

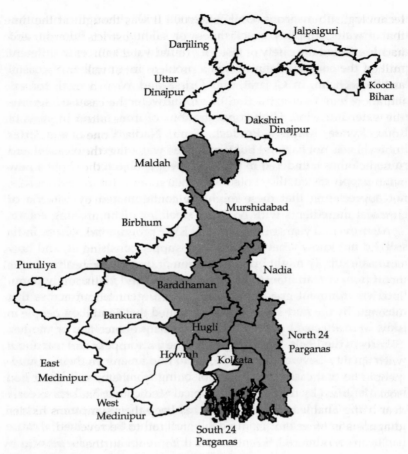

Map 7.2: Arsenic Affected Areas of West Bengal
Source: Government of West Bengal, Random Sample Survey, June 2002.

In West Bengal, the situation began to be addressed earlier and the problem was never as widespread. But by the late 1990s it was causing considerable concern, not the least because it threatened—as in Bangladesh—to become rapidly worse if not given due attention. In 1999, Unicef embarked with the government of West Bengal on a US$ 3 million Plan of Action to address the problem in a comprehensive way: epidemiological mapping, water surveillance, treatment of cases of illness, and the development of appropriate technological responses (UNICEF 2000). It was thought at the time that one-third of the water sources in the eight districts where arsenic had been found could be affected (UNICEF 2000). The entire 5.1 million people living in these districts were not at risk, but as many as 1.5 million might be, and more could become so in a far-from-static situation. This has subsequently proved to be the case. To discover the extent of the problem and establish where alternative water sources were needed, it was necessary to test each one of the 150,000 tubewells in these districts, and to keep testing them over time. Arsenic mitigation was not going to be easy, even if the problem was relatively confined.

Chronic exposure to arsenic, even via the ingestion of quite a small amount over time,[2] results in a condition known as 'arsenicosis'. The symptoms include skin lesions characterized by dark spots on the body, thickening of the palms and soles, flushing of the face, conjunctivitis, chronic cough, enlargement of the liver, and peripheral neuropathy. If the consumption of toxic water continues, skin and internal cancers can develop, as can other complications such as ulcers, cardiovascular diseases, and gangrene. The health effects of slow arsenic poisoning do not make their presence felt for a while. Skin lesions are usually the first signs and only appear after five years, with the more serious conditions developing many years later. Many people with the symptoms do not admit to them, shrinking from social stigma. Girls with the dark, tell-tale discolourations of arsenicosis fear being unable to marry. Since they keep their symptoms hidden, diagnosis is often too late for the condition to be reversed.

By far the best remedial action is to avoid further exposure by drinking arsenic-free water. Alternative water sources were, therefore, a priority. Various options were possible for West Bengal. One was to drill tubewells in risky areas to a depth of over 100 metres, where they tapped a deeper, arsenic-free, aquifer. However, if this was done without great care, contamination might seep downwards. Another possibility was to harvest rainwater from rooftops into household storage tanks, or use other sources of surface water filtered to

---

[2] The Indian government regards 50 ppb as the maximum permissible limit of arsenic in the water supply. However, over a very long period, this level is still toxic. The WHO recommends 10 ppb as the maximum permissible limit; this is so small an amount it is difficult to trace accurately.

remove bacteria. Various forms of large filter, using sand or gravel packs, can be attached to ponds and tanks, and are technologically simple, cheap and effective. However, the degree of community cooperation and organization required to keep them clean and functional makes such systems of filtration difficult.

In West Bengal, the field test kit developed by the AIIHPH, Kolkata, was at the heart of the proposed water surveillance strategy. The kit could include not just a bacteriological test, but reagents for testing arsenic and fluoride as well. In any water sample, arsenic could be detected to the level of 50 ppb (parts per billion), and the result obtained in only 15 minutes. However, in 1997–8, Unicef commissioned the Shriram Institute for Industrial Research in Delhi to undertake a thorough assessment of all water testing field kits then available in India, with a view to establishing the credentials of the best available. Although the AIIHPH kits came out better than many, the upshot of the assessment was that none were adequate. Therefore, Unicef invited the National Chemical Laboratory in Pune, Maharashtra, to develop a more accurate, less commercially-driven, and more user-friendly portable field testing kit (Shriram Institute for Industrial Research 1998). It would also be more sensitive. For arsenic testing, it should be able to detect levels down to 10 ppb.

Unicef also provided the district laboratories at the second tier of the evolving West Bengal community water quality surveillance system with specialized equipment: spectrophotometers. At the third and highest tier, 5 per cent of all samples were assessed at laboratories in Kolkata, using highly sensitive atomic absorption spectrophotometers measuring to 5 ppb. Unicef and the government of West Bengal are gradually putting in place a system which deploys community-level field tests at its leading edge, backed up by the use of spectrophotometers to check out the results and ensure accurate surveillance. The process of creating a supply chain of testing and cross-checking for all potentially-affected water supply sources is still undergoing experimentation. In time, it is hoped that marketing water quality tests to the owners of private tubewells, of which there are almost ten times as many as public tubewells in West Bengal, will allow the testing system to be self-supporting.

A great deal of effort surrounding the issue of arsenic in drinking water in West Bengal and in Bangladesh has gone into discussion of the causes and establishing the epidemiology of the problem. Once a handpump supply is found to contain arsenic, the pump is painted red to warn people not to use it for drinking or cooking, only for laundry and cleaning. But it may be some time before alternative drinking sources are provided. What people need is an immediate remedy—something they can apply instantly without reference to the authorities. Unicef has, therefore, been helping to develop a low cost household filter. The device is similar to the standard ceramic double chamber water filter used to remove excess iron and bacteria, with an additional chamber containing an activated alumina 'sachet' or pouch. This is attached internally in such a way that the water must percolate through it.

The filters are made by trained artisans and there is already great demand for the product. They will eventually retail at less than Rs 200 (US$ 4). But although Unicef has been working on the design for some time, and is as enthusiastic as anyone to get the filters out in the community, it is proceeding with caution. There are many risks with such a toxic material as arsenic. A number of questions regarding the device remain unanswered. A primary one is how long can the activated alumina pouch be trusted to work effectively? It will need to be reactivated after a certain number of months, and after a certain number of reactivations, the arsenic waste sludge will need to be safely disposed of. The sludge too, needs safeguarding. As a first step, 60 households have been supplied with filters and are being regularly monitored. Over a period of eight months to two years, it will be possible to see how long the pouch remains active, and what conditions of water temperature and turbidity affect the filters' performance. No such product should undergo mass release until its own quality is assured, and necessary systems associated with it—reactivation, sludge disposal, pouch replacement—fully in place.

The arsenic problem in West Bengal and Bangladesh is the drinking water quality issue which has attracted notoriety from both inside and outside the Indian sub-continent. But the problem of fluorosis is much more widespread, affecting no less than 62 million Indian people, a large proportion of whom are children (Susheela 1999).

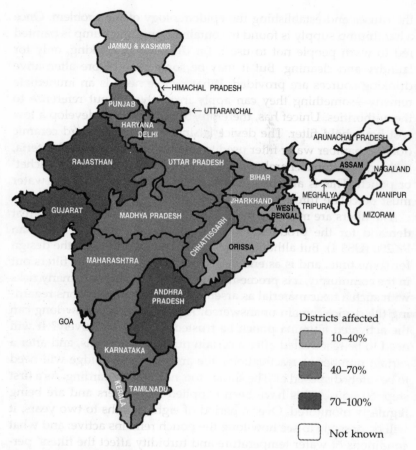

Map 7.3: Fluoride Affected States in India

Source: State of Art Report on the Extent of Fluoride in Drinking Water and resulting Endemicity in India by Fluorosis Research & Rural Development Foundation, New Delhi, 1999.

In the case of arsenic, so far only very limited areas of the country are seriously affected. But in the case of excess fluoride, the groundwater is affected in every district of Andhra Pradesh and Rajasthan, and in large parts of several other states. Like arsenic, fluoride is tasteless, colourless, and odourless. Ingested over a long period it affects multiple tissues, organs, and systems of the body and results in a crippling condition of the bones or soft tissues. In children and young people whose teeth are not fully developed, it also leads to dental damage and discolouration. Dental and skeletal fluorosis are impossible to reverse. The treatment for soft tissue fluorosis—aches, pains, and stomach disorders—is a change in dietary habits and fluoride-free drinking water. Early detection and prevention are, therefore, vital.

Of all the affected states, Rajasthan suffers worst. Over 70 per cent of the population of Rajasthan is dependent on groundwater for drinking purposes, a proportional increase since the days of guinea worm eradication and the discouragement of step-wells, ponds and other open sources for water use and collection (Jamwal and Manisha 2003). However, in the case of 20 per cent of rural dwellers, the groundwater they now use because it is supposedly safe from guinea worm and other infectious agents is by no means 'safe' as far as fluoride is concerned. A survey of habitations conducted in 1991–3 by the PHED found that out of 37,889 villages in the state, 9741 had more than the danger level of 1.5 ppm (part per million) of fluoride in groundwater, and of these 3280 have more than 3.0 ppm (*Hindustan Times* 2001). In 1999, Rajasthan became one of four states in which Unicef helped support a major push against fluorosis in partnership with the government and local NGOs. Public awareness is a high priority, together with promoting other sources of drinking water, changes in diet, and defluoridation of water by using household filters.

In the village of Choriya, in Tonk district, excess fluoride has been a problem for many years. Tests carried out on the four handpumps used by the 72 families in the village showed that they had between 8.9 ppm and 11.4 ppm of fluoride—an amount way above the danger level. But until they were informed why they were so frequently sick, people in the community did not realize that the water supply was to blame. Goji Mina, the sarpanch, now 60 years old, can remember that when he was a young man of 20 there was no problem with the water here. 'We used a dug well for drinking in those days. But the water table dropped and we were forced to use another source. In 1972, the PHED brought boreholes and handpumps. After that we had problems, but we did not make the connection. When someone had indigestion or aches and pains, they went to the doctor to get treatment. Many people paid for treatment which was useless.' Light was shed in Choriya only two years ago on the reasons for bent bodies, buckled limbs, and damaged teeth. Some other Rajasthani villages—Jharana Khurd at a distance of only 20 kilometres from Jaipur, for example—are in a far worse state. Here, all 1200 people irrespective of age look old, have cracked teeth, and perpetually aching shoulders, hips, and ankles. All 25 tubewells bar one have dried up in the past three years because of the drought, and the water from the remaining one is 'poison' (Jamwal and Manisha 2003).

Choriya village is now part of a pilot project to reduce the fluoride content of water by use of household filtration techniques. A local NGO—the Centre for Community Economic and Development Consultant Society or Cecoedecon—is operating the project in eight villages, using two different defluoridation systems. In Choriya, the system being tried is a household bucket filter, using activated alumina granules through which the water passes. But the system does not simply rely on distributing the filters to select members of the community. A *pani panchayat* or water committee has been elected to look at all the matters relating to water in the village, an animator has been trained, water points mapped and tested, handpump user groups set up, rallies and village contact drives undertaken along with health education. The beneficiaries for filters have been chosen with the full participation of the village. This is an inclusive water and sanitation programme, in which the anti-fluoride activity is at the core.

One of the priorities was information and treatment for fluorosis sufferers. Shikhar, the NGO facilitator, describes the dietary changes they propose. 'More citrus fruits, vitamin C, curd, less of tea, less of smoking, and no *paan* (a digestive containing betel nut). When they start consuming treated water and eating differently, appetite improves and the pain in their legs goes away. There is relief almost immediately. This makes them highly motivated.' However, rarely enough for them to abandon smoking and paan chewing. 'We also hold medical camps where we check for dental and skeletal fluorosis. This helps build awareness. It is still difficult for some people to believe that their children's legs are bent or their teeth damaged because of the water supply. They don't expect such things to have this kind of explanation.'

However, the results of household filtration do carry conviction. Altogether, 24 stainless steel bucket filters are in use. The materials for water testing and reactivating the alumina granules are kept by the animator in a small depot in the village. She visits the 24 families regularly to test their treated water, makes sure they are using the filter correctly, recommends when to bring the granules in for activation, and motivates them with regard to diet. A small charge of Rs 22 is made for reactivation of the granules, which takes about eight hours. This needs to be done around every three or four months, and as many as 20 reactivations of the same materials are possible.

All the units in Choriya are in use. As many as 73 per cent have a safe water output, and 84 per cent of the people using them report a positive change in their own and their families' health (Cecoedecon 2002). At present, the filters, costing Rs 600, are too expensive for poor families to buy. But prices will go down, and systems for subsidy or loans for families below the poverty line can be introduced, as they are in other contexts. The pani panchayat has collected Rs 5000 (US$ 100) for other water-related activities, which might include fluoride control in due course but at present is more likely to be spent on drought mitigation. Already, the price of alumina granules has dropped from Rs 300 for a 3 kg bag, to Rs 250. If prices continue to drop, it will not be many years before the technology for fluoride mitigation can be absorbed into the local economy of Rajasthan.

In Anantpur district in Andhra Pradesh, the state with the second highest fluoride problem in the country, this has already begun to happen. Here too, Unicef worked in cooperation with the government authorities, and with an NGO—MYTRY Social Services Society—an organization working with marginalized groups in Anantpur for many years (UNICEF 2001). In the 10 blocks where MYTRY was operational, the project strategy was based on the introduction of Domestic Defluoridation Units, developed at Unicef's initiative by the Indian Institute of Technology, Kanpur.

In the course of two years, 1950 units were distributed in 31 villages at a nominal price of Rs 250 for those below the poverty line, and Rs 400 for those above it. A further 7600 were sold by MYTRY's Rural Sanitary Mart. Village Fluoride Volunteers perform the same task as animators in Rajasthan, visiting households, providing information, monitoring water samples, and carrying out alumina regeneration.

They also act as sales agents for the defluoridation units, for which they receive a commission. So popular have the units been that an NGO from an adjoining neighbourhood ordered 2000 from MYTRY. Entrepreneurs have joined in, and the anti-fluoride effort has become entirely consumer-driven, with little intervention by the government. However, this cause for celebration needs to be tempered by the realization that, without quality control, risks can arise with sub-standard filters. Here, once more, is a case where manufacture is being stimulated more by the prospects of profit than of disease control.

Water quality surveillance has been a relative latecomer to Unicef's water and sanitation programme priorities. So much attention was paid to bacteriologically 'safe' water in the early decades of the programme that the need to test for chemical contamination did not receive the attention it should have. However, now that Unicef is addressing this issue with government partners, the important feature of its approaches is that they are not exclusively based on technology and 'fixing' the problem by means of fool-proof filtration, placed in the hands of centralized professionals, with all the staff and overhead costs that this type of approach entails. The social dimensions of the intervention have been given as much attention as the technical, and the whole thrust of the effort is to enable people to deal with their own problems of fluoride, arsenic, iron, or salinity with as little dependence on the government as possible. In these multi-partner approaches with the government, NGOs, and small entrepreneurs, and in its determination to build local awareness and capacity, are many of the lessons Unicef has learnt over the 30 years of its programmatic involvement with water in India.

However, there is one area which local people cannot control. This is the extraordinary pressures to which the groundwater reserves are being subjected, the stresses imposed by over-extraction from aquifers, the lack of recharge, and the spiralling disaster of declining water tables, increased contamination, and crop failure at village level. Or can they?

In the early 1980s, the village of Ralegaon Siddhi in the drought-prone area of Maharashtra was in a state of destitution. Less than 5

per cent of the poor-quality agricultural land was irrigated and much of it had been mortgaged to moneylenders. There was little work in the village. Around 20 per cent of families ate just once a day. Many of the rest borrowed grain from other villages at high cost and became increasingly indebted. With no other source of income, people had taken to manufacturing liquor. There were 35 to 40 liquor stills. Drunkenness was common; and with it came feuds and crime, specially against women, some of whom resorted to prostitution. A few village wells were useable. But these were major sources of disease— 90 per cent of the families were stricken by stomach problems. Child mortality was high (Shourie 2003).

This village has since enjoyed an extraordinary renaissance, and become a place of pilgrimage for those who want to witness at first hand what can be achieved by rainwater harvesting. Today, all available water is systematically harvested by percolation tanks, check dams, bunding of nullahs, contour bunding, additional dug wells, aquifer recharge, and a whole gamut of water harvesting techniques. Around 1200 acres—75 per cent—of the arable land is sculpted and water-fed by the system. Three crops are grown, and vegetables, grains, and milk worth Rs 5 million are marketed to major cities every year. Around 200,000 trees have been planted. While neighbouring villages wait for government tankers to bring drinking water, Ralegaon has enough for everyone in the village and to spare. The cost of the original investment in structures was no more than Rs 2 million (US$ 40,000), part provided locally, part by the government. All the work and effort has come from the community. So apparently, even in an unforgiving environment, people can reverse the disastrous water crisis which is eroding the ways of life for many communities.

The instigator of this miracle is Anna Hazare, a follower of Swami Vivekananda and of Gandhi, who took early retirement from the army to return to his village 17 years ago and try to turn its fortunes around. He studied the topography surrounding Ralegaon, and developed watersheds by checking the run-off from sudden seasonal downpours, thus conserving both soil and water. Hazare's achievements are truly spectacular on many levels. There is no drinking or smoking in Ralegaon, there are no child marriages, all girls are in school, there are no dowries, and there is social integration between high caste and *dalit* families. All this is a result of his conviction that everything depends on community effort, and the way he painstakingly encouraged that effort and proved to everyone's satisfaction the value of mutual aid. However, the extraordinary role and influence of Hazare on what has happened in this village, although it proves what can be done with a potent combination of rainwater harvesting and community action, means that the accomplishment is not easily replicable. This is a familiar feature of truly effective community transformation, and it is, therefore, a feature of many rainwater harvesting endeavours.

In 1997, the Centre for Science and Environment (CSE) in New Delhi published a seminal report in its 'State of India's Environment' series: *Dying Wisdom: Rise, Fall and Potential of India's Traditional Water Harvesting Systems*. The report was the accumulation of 10 years of work, led by the late Anil Agarwal and Sunita Narain, two of India's most prominent environmental activists. Two impulses led to an almost missionary focus by CSE on rainwater harvesting. One was the growing anti-dam movement in the country, which begged the question of finding less socially and environmentally damaging alternative means of water retention and storage. The other was the discovery in 1987 in a remote corner of the Thar desert of unique traditional water devices—*kunds* or carefully constructed saucer-shaped catchments—for collecting and storing rainwater. These were allowing people to meet their water needs even at the height of a devastating drought (Agarwal and Narain 1997). Agarwal and his team became aware that, even with a very small amount of rainfall annually, a community that 'captured the rain' could provide for all its water needs without difficulty, and at the same time could reverse the declining water table trend in its immediate environs.

Subsequently, the CSE undertook a quest to rediscover the various surface and groundwater harvesting techniques which had served India's different environments so well down the centuries—until the colonial engineers of the late nineteenth and early twentieth centuries, and their Indian inheritors, let them go to rack and ruin. The wealth of evidence of the effectiveness of these many traditional types of systems in meeting local needs was breathtaking. And the assertion that communities could, with relatively little investment, re-green their environment, replenish the local water table, and transform their

Thunti Kankasiya village	before 1991	1999
Perennial drinking water wells	Nil	23
River dams	Nil	1
Months of water availability	4	12
Land under cultivation (hectares)	85	135
Number of crops per year	0–1	2–3
Agricultural production (quintal/hectare)	900	4000
Migration rate	78%	5%
Average period of migration (months)	10	2
Income per household (Rupees per year)	8590	35,620

FIGURE 7.2: An example of micro-watershed development in a village in Gujarat

Source: Sadguru Water and Development Foundation, Dahod, Gujarat; quoted in Down to Earth magazine, 15 January 2000.

farming fortunes has since attracted widespread notice, including from many figures of national and international importance.

The attention captured by CSE's campaign for rainwater harvesting as a response to the country's water crisis—a response which, until a few years ago, was seen as the preoccupation of a few eccentrics—has augmented the flurry of VIPs and development pilgrims who now go to inspect the work of Anna Hazare and similar practitioners in Maharashtra, Rajasthan, Gujarat, and Madhya Pradesh. Another well-known and much garlanded example of water harvesting success is the work of Rajendra Singh and the Tarun Bharat Sangh (TBS) in Alwar district, Rajasthan. In the basin of the Arvari river, the landscape has been entirely regenerated by the construction, over several years, of hundreds of bunds, johads (small earthen dams), and other structures. Instead of a barren and deserted valley, there are now thriving agricultural communities. Each village has played a central part in the process, identifying the familiar paths of streams, repairing old johads, contributing 30 per cent of the costs of any new construction and all the unskilled labour. People have become increasingly committed as the benefits of re-greening became evident. The 72 villages in the valley have developed their own system of water government—a 'water parliament'. This meets once a month on Black

Moon day,[3] which they all take leave to attend. The communities have made their own rules about water extractions, crops which may and may not be grown, and about preserving trees. Infringements are met with fines: Rs 20 for cutting down a tree, and Rs 50 for not reporting someone who cut down a tree, and so on.

There are other examples of extraordinary success in regenerating defunct wells, tanks, and waterways, in parts of Gujarat, in various districts of Maharashtra, and in areas of Karnataka, Andhra Pradesh, and Tamil Nadu. In many cases, these have come about because of the dedicated leadership of key individuals, and because of their belief in community-based action and community resource management. But, in India's political climate of dependency on government schemes and grand supply-driven interventions, this is as much a cause for difficulty as for celebration. Some of the gurus of water harvesting face official obstruction, and are looked at askance for their radical ideological views about people's capacities to defend their own resource base and generate their own development. TBS leader Rajendra Singh has frequently been harassed, even physically attacked. He has been taken to court over 300 times for building illegal structures under legislation left over from the British Raj.

As long as there is no awkwardness with traditional land and water rights—a typical basis for obstruction—the actual building of water harvesting structures is relatively easy. This is yet another construction task which any contractor can undertake, given a little expertise from someone with a hydrological and topographical grasp of the local terrain. But building a social structure around the physical structure which can manage the installation on behalf of the community, equitably, efficiently, and sustainably—that is a much more difficult task. As Anil Agarwal of the CSE pointed out to an audience of members of Parliament and state legislatures in 2001, 'Water is a strange natural resource: it can unite a community as easily as it can divide it. Therefore, it is essential that a strong social process precede each structure to build what economists call the 'social capital'. This is an area where the track record of government agencies is literally

[3] Black Moon day is the day in every month between the waning of the old moon and the new moon's appearance, and is traditionally a day on which community business is done in this part of Rajasthan.

non-existent, and inflexible government rules militate against the very principle of social mobilization' (Agarwal 2001). Unicef would not put it as strongly as this. But its own experience in trying to develop systems of community maintenance for drinking water supply schemes illustrates that, without NGO intermediaries doing the social mobilization for them, government bodies could point to few, if any, examples where they have prompted truly participatory, community-led development.

In 1995–6, Unicef embarked on its own attempt, in partnership with the World Wide Fund for Nature (WWF), to establish the seriousness of India's freshwater crisis and outline some principles for response. The rationale for Unicef's concern was that, after all the investment and effort that had gone into providing children and families with safe water and sanitation over the previous 30 years, the achievements, with which it was proud to be associated, were being put at risk by poor water management and lack of environmental protection. The major problem was the lowering of the water table and the fact that many boreholes were running dry.

In spite of expressions of official concern over watershed deterioration since the mid-1980s, very few steps had been taken to stem deforestation, recharge aquifers, and introduce regimes for watershed management. The effects, as always, were most severely felt among the most vulnerable members of society, including the marginalized, the landless, the poor, and women and children. Some panchayats confronted by wells and boreholes with no water in them abandoned their handpumps, deepened the boreholes, and installed electric pumps for lifting water to the surface. But poorer communities and poorer sections of communities could not afford to do such things. The failure of policies at the macro-level to control freshwater extraction and manage water resources in an efficient and integrated way was jeopardizing the future health and well-being of millions of children.

Unicef and WWF jointly undertook 11 studies of how people in water-stressed environments were coping with the threats to health, livelihoods, and local ecosystems. The studies were on a small scale, and were conducted in different ecological settings; the arid, the drought-prone, the mountainous, and the highlands. They were significant in recognizing that local parameters—of hydrogeology, agricultural patterns, socio-economic livelihoods, and cultural dynamics—should be the mainspring of policies designed to deal with water problems in any given setting. For the first two decades of Unicef assistance to water supply and sanitation in India, there had been a broad acceptance of the idea of centrally planned and directed programmes, whose strength had lain in streamlining technical inputs and controlling costs and quality. But in the context of sanitation, and of community maintenance of water supplies, where the performance of social as well as technical systems was critical, trying to develop universalist prescriptions and implement them had proved unworkable.

There remained—and remains—some organizational reluctance to accept that the quintessential feature of any truly participatory approach is that it has to be flexible to local circumstances and is irreconcilable with the notion of a programmatic template. But the engagement with WWF and the broadening of the analysis to include the local ecosystem, prompted Unicef to move towards a different vision in water and environmental sanitation. An understanding of what was happening in a given locality, of what people knew about local freshwater dynamics, and how they coped with their specific problems, would provide the starting point for sound local water resource management. Thus each of the studies, which were conducted by local institutions and NGOs, strove to work out the particularities of water problems and solutions in each situation by means of data collection and by clarifying local perspectives from exercises in participatory rural appraisal.

The report on these joint studies was issued as *Fresh Water for India's Children and Nature* by Unicef and the WWF in Delhi in 1998. It was intended by its Unicef co-authors to propel the water supply programme in India in a new direction, to lead to a preoccupation with freshwater management policies as part of the organization's ongoing commitment to the well-being of future generations of Indian children. A number of policy recommendations were included, which underlined the need for macro-level changes in legislation, regulation, policy development, and action, but also recognized that these in themselves could not transform the situation at the micro-level. Here, many options, including rainwater harvesting, water-sharing arrangements, and local water markets could be effective in helping to restore depleted fresh water sources.

But for various reasons, this step in the direction of a new Unicef involvement in water resources management was not taken. For one thing, there was no responding chord of resonance from the government as had been the case earlier with boreholes, handpumps, school sanitation, and water quality testing. These were the components in the Unicef portfolio associated with technology, equipment, and construction—the perennial favourites of bureaucrats and politicians. But the serious control of groundwater extraction, disincentives for growing water-guzzling crops in water-short areas, and the reduction of subsidies for electricity used to pump irrigation water are all political hornets' nests. And though there are signs in enlightened corners of state and district machinery of real enthusiasm for small-scale, community-based solutions to water conservation and management, as yet these have not won widespread official recognition as a substitute for much grander engineering schemes. On the contrary, current evidence tends the other way (see Chapter 8).

Unicef also had its own organizational reasons for not entering such a broad policy area at this time. Attention to water supply, as opposed to sanitation and hygiene, was receiving lower programme priority. And even in the heyday of support for water supply, Unicef had never conceptually linked the use of water in the home and for drinking, and its use for livelihoods and food security. Water and sanitation had always been seen by Unicef as public health interventions, not as livelihood interventions, never mind perceptions at the community level. Now that there was some fusion of health and livelihood concerns in an overall thrust for integrated water resources management, Unicef, after a brief flirtation, backed out. A concern with overall community water resource management, as opposed to management of drinking water, was perceived as too distracting at a time when the organization was trying to narrow the focus of its assistance to areas which directly addressed the health and well-being of children.

However, the deepening water crisis in India, and the need to address issues of water resources at the community level in the interests of long-term sustainability of the livelihood base, can never recede. On the contrary, they worsen year by year. In various ways, over the following few years, Unicef's programme of support for water and environmental sanitation was bound to find itself thrust into an engagement with livelihood as well as public health issues. While the days of large-scale famine may have disappeared, there is still chronic hunger in many marginal communities—hunger exacerbated by drought, crop failure, livestock and livelihood loss. This is not an issue from which any humanitarian organization committed to the well-being of children can ultimately stand aside.

REFERENCES

Agarwal, Anil (2001), *Drought? Try capturing the rain*, briefing paper for members of parliament and state legislatures, occasional paper, CSE, New Delhi.

———— and Sunita Narain (eds) (1997), *Introduction to Dying Wisdom: Rise, Fall and Potential of India's Traditional Water Harvesting Systems*, CSE.

All India Institute of Hygiene and Public Health (1993), *Water Quality Surveillance Report*, quoted in *Medinipur Shows the Way: Water Quality Surveillance: a Community-based Approach*, UNICEF, Kolkata.

Black, Maggie (1990), *From Handpumps to Health*, UNICEF, New York.

Cecoedecon (2002), materials provided during project visit in October.

Chadha, D. K. and Santosh Kumar Sharma (2001) *Central Ground Water Authority: A Vehicle to implement Ground Water Law in the State*, GSDA, Government of Maharashtra, September.

Government of India (2002), National Water Policy, Ministry of Water Resources, April.

———— (1994), National Register of Large Dams—Central Water Commission, New Delhi.

Hindustan Times (2001), 'Fight Fluorosis and Save our Children', Supplement on fluorosis prepared by the Hindustan Times, and re-published by UNICEF Jaipur and the Government of Rajasthan, 2001.

Jamwal, Nidhi and D. B. Manisha (2003), 'Wah India', Cover Story in *Down to Earth* magazine, CSE, April.

Nigam, Ashok, Biksham Gujja, Jayanta Bandyopadhyay, and Rupert Talbot (1998), *Freshwater for India's Children and Nature*, WWF and UNICEF.

Panda, Ramesh Chandra (2001), *Integrated Approach for Sustainability of Drinking Water Supply*, GSDA, Government of Maharashtra, September.

Saxena, N. C. (1999), 'Rural Water Supply', paper delivered during State Water Ministers' Workshop on Rural Water Supply Policy Reforms, Cochin, 7 and 8 December.

Shourie, Arun (2003), 'Gandhi is alive in Ralegaon Siddhi: Meet Shri Anna Hazare', NRI-homecoming.com/MV RalegaonSiddhi.html.

Shiram Institute for Industrial Research (1998), *Comprehensive Report on Assessment of Water Testing Field Kits*, 1997–8, Delhi.

Susheela, A. K. (1999), 'Fluorosis management programme in India', *Current Science*, Vol. 77, No. 10, 25 November.

UNICEF (2001), 'Combating Fluorosis', Andhra Pradesh, Child's Environment Programme, information note, UNICEF India, draft.

———— (2000), 'Arsenic contamination in India', information brief, UNICEF India.

———— (2000), *Learning from Experience: Evaluation of UNICEF's Water and Environmental Sanitation Programme in India, 1966–98*, Evaluation Office, UNICEF, New York.

———— (1996), *Medinipur Shows the Way: Water Quality Surveillance: a Community-based Approach*, UNICEF, Kolkata

———— (1996), Case study on 'Community-based Water Quality Surveillance and Monitoring under the CDD–Watsan Project', Allahabad, Uttar Pradesh, UNICEF, Lucknow.

———— (1995), Watsan India 2000, UNICEF, Delhi.

8

Water, Life, and Health:
Where Next?

Every year towards the end of May, when the temperature reaches
its unbearable zenith throughout the sub-continent, the people of
India hold their collective breath for the arrival of the monsoon. The
due date is 1 June. The rains usually sweep into Kerala on time from
the Indian Ocean for their annual visitation, though there is concern
in some quarters that they are becoming less punctual and behaving
more erratically than they used to. Once the clouds have burst over
Kanniyakumari (Cape Comorin), the stormy deluge moves rapidly
north. Every early downpour, never mind flooded streets, power-
cuts, falling branches, minor damage and inconvenience, is accompanied by joyful celebration—the yearly renewal of life throughout
the country has begun. Children splash in puddles on their way to
school. Farming families watch with relief as streams and wells fill
up, trees burst into leaf, seeds germinate and sprout. Mothers and
health workers look out for the coughs, colds, and diarrhoeal infections which show that all living organisms are enjoying a rebirth now
that the wet season has arrived.

The rains advance in their familiar pattern: up the Western Ghats
towards Goa, Maharashtra, and Madhya Pradesh; north-east into
Andhra Pradesh, Bihar, Orissa, West Bengal, and Sikkim. By late June
the central part of the country is covered, and by early July, the
western and north-western states—Haryana, Uttar Pradesh, Delhi,
Himachal Pradesh—have received their first showers. In the far west

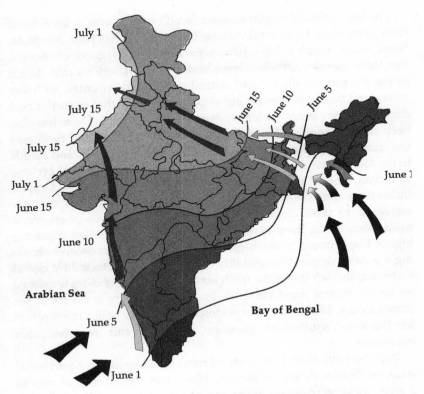

July 1

July 15

June 15

June 10

June 5

July 15

July 15

July 1

June 15

June 10

June 1

Arabian Sea

June 5

June 1

Bay of Bengal

Map 8.1: Normal Dates of Onset of the South-west Monsoon
Source: Raj Kumar Daw, data from mapsofindia.com

of Gujarat and Rajasthan, in the Rann of Kachchh, and in the Thar Desert at Barmer and Jaisalmer, these may be the only showers of the year. The urgency which attends the coming of the monsoon is ever deeper in parts of the country—Saurashtra in Gujarat and parts of Madhya Pradesh—where water shortages have recently been so acute that, in the pre-monsoon weeks, violence has erupted at distribution points (Athavale 2003). The meteorologists are in constant demand for forecasts and predictions, and in drought-prone states such as Karnataka, Maharashtra, Andhra Pradesh, Gujarat, Rajasthan, and elsewhere, drought monitoring cells and hydrological institutes work flat out to produce detailed rainfall data.

The last general drought disaster faced by India was in 1987. Since then, rainfall, as far as total volume is concerned, has been adequate. Some years, notably 2003, have seen reassuringly good monsoons. But there have nonetheless been many years of serious rain deficit in the western, southern, and central states of the country, with key rivers and reservoirs becoming depleted at rates that even the odd excellent year cannot replenish over the longer term. There has also been some evidence of extra volatility in the pattern and timing of the monsoon, and a number of extreme weather events and floods in the North-East and on the eastern coast, including the 1999 cyclone in Orissa, one of the worst ever weather-related disasters. Despite indications that the intensity of floods and droughts has recently increased in those parts of the country where these phenomena most often occur, there is official resistance by the Indian Meteorological Department to the suggestion that India's monsoon is showing a greater degree of variability than usual or is affected by global warming. But whatever the truth about the impact of climate change on temperatures, wind currents, and rainfall, there is no room for complacency. Every year sees widespread anxiety about the forecast for the main south-west monsoon, its onset, and its subsequent behaviour.

Will the pattern and intensity of rains be adequate in a particular state, or district, or micro-climate? How does the volume of rain in a given location compare at the end of July or the middle of August to the norm (plus or minus 20 per cent of the average)? Will the rains withdraw and peter out too early in the north-western desert regions? Or have they arrived with such a tempestuous deluge in the east that devastating floods will follow, inundating crops, bursting river banks, threatening dams and sluice-gates in such a way that precious run-off will have to be allowed to flow unchecked into the sea? Such a flood disaster caused devastation in Orissa in 2001, with 600,000 villages marooned, 550,000 hectares of standing crops destroyed, and seven million people affected (Shiva 2002). Although the north-east monsoon lands heavy rainfall on Tamil Nadu and adjacent parts of Kerala from October to December, and provides many western and northern states with enough rain to plant a smaller second crop, the period of the south-west monsoon between early June and early September is critical. Ultimately the fortunes of the farming economy, the economic and social well-being of every rural household, and the ecological health of the natural resource base via the refilling of aquifers and streams are all in thrall to this meteorological event. In India, everything depends on the monsoon—as this book began by observing.

This has been the case since time immemorial, and after 55 years of independence and more than two generations of 'development', little has altered on the surface of things. Even though agriculture plays a less prominent role in the country's economy than it used to, accounting for just over a quarter of the country's gross domestic product (GDP) instead of around two-fifths as was the case in the 1970s, 70 per cent of the country's people still live on the land. Their livelihoods and well-being derive directly or indirectly from crop cultivation and livestock. The condition of these in turn depends on the reliability of water sources, whether they exclusively look to the sky for rain or also use some form of irrigation. Many towns and cities also experience serious shortages as the dry season progresses. Rajkot in Gujarat, Delhi, Chennai, Mumbai, and Hyderabad, for example, are frequently in the news. These shortages cause most misery among the poorest communities.

But even if the monsoon seems to be an event of unchanging significance, the way in which the rains are interpreted within the context of Indian development policy has altered in recent decades. This stems partly from a different way of looking at water. Ever since the first Earth Summit in Rio de Janeiro in 1992, the international community has been promoting a view of water which is at odds both with traditional value systems and with the social welfarist attitudes towards basic needs and essential services which prevailed in the 1950s, 1960s, and 1970s. Water is no longer seen as a free good, provided courtesy of the laws of nature, in which all members of humanity share rights in common. Instead, water is a precious and finite resource which should bear a realistic price tag. Although beliefs and attitudes among rural people in India and elsewhere remain unaltered, in the official perception, water's life- and health-giving functions have been coopted into the realm of economic and political affairs. Water is no longer seen as a god-given resource, but one over which human agency, for good or bad, holds sway. Rainfall is the basic ingredient of the resource, but as far as most planners are

220 Water: A Matter of Life and Health

concerned, where rain naturally falls or collects above and below ground is no longer the determining factor in how water should be managed or used. The new international 'vision' of water, and of the policies and practices which should guide its management and use, has been coopted by India in its own particular way.

This change in perception has been accompanied by an altogether more scientific and secular version of what water abundance and scarcity are about. Like other resources, water can be assessed. Its quantity and quality can be measured, and amounts of the resource can be allocated by the application of rules of supply and demand. If there is a shortage in one area, a transfer can be made from another by tanker or train—as regularly happens in Maharashtra, Tamil Nadu, Gujarat, and Rajasthan. Over the longer term, this can be done by pipeline, canal systems fed from reservoirs behind dams, or even by the linkage of rivers. A vast scheme to harness floodwaters from 14 rivers feeding into the northern reaches of the Ganga and Brahmaputra, lift and pump them over the Vindhya mountains, and use them to replenish 17 depleted water-courses of the south, was announced by the then prime minister, Atal Behari Vajpayee, in December 2002 (Pearce 2003). This massive engineering project, known as the 'garland of rivers', is far in spirit from the international vision of holistic water resource management, in which economic, social, environmental, financial, and equity concerns would be integrated within river basins and catchments. Transfers between basins have become almost a leitmotif of current Indian water policy, which in turn is based on water-intensive irrigation requirements for Green Revolution farming approaches in parts of the country where the average rainfall is insufficient to support them unaided (Shah 1993).

In India, as in other parts of the world, water shortages, and the distances water supplies have to be transported, have added enormously to the value of the raw resource and to the costs of water-related infrastructures and services. Their expenses have a cyclical and exponentially rising character: hydropower is needed for electricity; much of this electricity is needed to pump water for irrigation and expanding municipal requirements, further exhausting the resource. And since the development, management, and use of water resources are dependent on ever larger-scale financial inputs and transactions, the resource has also emerged as a tradable commodity—one whose price tag rises higher along with its scarcity, and around which entrepreneurial speculation and profit-making are bound to increase. This is another potential source of conflict. Protests recently erupted in Chhattisgarh over the leasing of a stretch of the Shivnath river to a private company, thereby excluding from its use local fishermen who fish its waters, and farmers who irrigate their vegetable crops along the banks. This is one of the many examples of the impacts on rural people of the commodification of water, and the gains and pains now associated with the manipulation of this resource for economic development. The 'garland of rivers' project, around which controversy is acute, is expected to cost up to US$ 200 billion, suggesting, as critics such as the CSE have pointed out, a contractors' paradise.

In the past four decades since Unicef first became involved in a programme of borehole drilling to support drought-relief efforts, even the idiom in which monsoon-related calamities is depicted has changed. One difference is in the ascription of cause. Few Indian environmentalists today accept the notion of 'natural' calamity in the context of water-related crisis. Man-made dynamics—the reduction of forest and vegetation cover, population growth, pollution of water bodies, rapid industrialization and urbanization, expanding consumerist lifestyles and other 'development' processes—have made a major contribution (Shiva 2002). Paradoxically, the Green Revolution, which enabled the country to achieve a state of agricultural productivity that nature's whims could no longer threaten, is contributing to the problem. It has imposed immense pressures on land and water resources, introduced problems of waterlogging and salinization in such key farming areas as Punjab and Karnataka, and been accompanied by the inefficient use of inorganic fertilizers and pesticides, which add further to river pollution loads (Lal 2002).

Mismanagement and misuse of the water resource itself are, therefore, integral to the nature of the growing crisis surrounding water in India. Farmers in drought-prone states such as Karnataka, Maharashtra, and Gujarat are unable to resist the incomes to be made from high water consuming crops such as rice and sugar cane. In Maharashtra, sugar cane may occupy only 4 per cent of the irrigated area, but it consumes 75 per cent of water used for irrigation (Murthy 2000). In Gujarat, studies show that when new canals are opened up for irrigation, instead of maintaining existing cropping patterns, farmers in the canal command zone shift to water-intensive cash crops such as sugar cane, exacerbating rather than reducing the overall water deficit (Government of Gujarat 2000). Thus, the crisis relating to water in India is itself a product of 'development', and questions about how to deal with it go to the heart of what the grand project of Indian development is and should be about, in the minds of officials, bureaucrats, politicians, corporations, financial institutions, and citizens of all affinities, classes and persuasions.

The crisis is presented in scientific, technical, and economic terms, of demand for a resource outstripping its supply. In this analysis, the language is of withdrawals, available and utilizable resources, rates of recharge, flows in rivers, pollution loads, freshwater degradation and so on. The macro-hydrology of the crisis is as follows. Every year, India receives around 4000 billion cubic metres (bcm) of rainfall, much of which evaporates or transpires, and the rest of which ends up in its soils, streams, and rivers. Since rivers fill at great speed in the rainy months and flow tempestuously, up to three-fourths of the precipitation ends in the sea. Therefore, of the water available by this process of natural recharge to surface water and underground aquifers, only 1086 bcm is utilizable annually, or something over 1000 cubic metres (m^3) per head (given the current population of one billion). This compares to an estimated availability of 6008 m^3 per head in 1947 (GOI 2002). In 1990, the total withdrawal or utilization for all uses was 518 bcm which, measured per capita, was something above the amount regarded as sustainable for life (500 m^3) for the then population. Both per capita use and the population are rapidly rising. Well before 2025, India is expected to reach a condition of water scarcity (less than 1000 m^3 available per person per year) (de Villiers 1999). By the year 2050, the country's total requirement on current consumption trends is estimated at 1422 bcm (GOI 2002). Apart from questions about how the volume of water available from renewable sources can be equitably distributed, where will the extra supply—380 bcm on the basis of current estimates, but a vast amount on any assessment—come from?

It will have to come from the rainfall—especially from that proportion of the rain which currently deluges down on people, buildings, roads, hillsides, forests, plains and soils; swells streams and rivers;

and pours out relatively unimpeded into the sea. So in the end everything comes back to the rainfall, whether from the small farmer's perspective or from that of the economic planners. The rainfall is finite, whatever the oscillating vagaries of its volume and pattern. The question is at what stage in its journey from sky, to earth, to soil, to aquifer, to river, to sea, should the rain be captured and by whom. And how that question is to be answered is shaping up to be the most critical one concerning India's water resources future.

The year 2002 was a year of below average rainfall for most of the country, with serious and cumulative deficit for some districts and states. In August 2002, Prime Minister Vajpayee stated that several parts of the country could anticipate a worse drought than the one in 1987. The affected states included the western and northern states of Rajasthan, Himachal Pradesh, Chhattisgarh, and Uttar Pradesh; Maharashtra, Karnataka, parts of Gujarat, Madhya Pradesh, and Orissa were also affected. According to the September 2002 statistics of the Indian Meteorological Office, only 15 out of 36 of the country's meteorological districts received normal rainfall, with 19 receiving between 20 per cent and 60 per cent less than normal, and two receiving more than 60 per cent less than normal. By May 2003, 60 million people had been rendered unemployed, 250 million had insufficient drinking water, and water in major reservoirs was reduced to 36 per cent of normal (Kang 2003).

Without doubt, the poor monsoon of 2002 helped to generate extra political impetus behind the 'garland of rivers' announcement in December. By January 2003, starvation deaths were beginning to be reported in corners of Rajasthan—in some parts of which no rain had been received since 1998 (Stuart 2003). In today's India, with large stocks of surplus grain, a robust emergency relief planning and distribution system in every state, and sufficient financial reserves at the centre to support states in crisis, such stories are thankfully few. Mass starvation death—even its threat—is reckoned to be a thing of the past. But, nonetheless, large numbers of people in drought-prone or other water crisis settings find their livelihoods, their livestock, and their own and their children's nutritional status drastically threatened.

As is to be expected, the victims of the creeping onslaught of drought in remote rural hinterlands and near-desert environments are among the poorest, often the landless, or the adivasi. By a painful irony, the good monsoon that marginal farmers pray for also spells ruin to those whose land is subsumed by major hydro projects: 3200 adivasi families in Maharashtra lost their homes, crops, and livelihoods to the 2003 monsoon floodwaters behind the Sardar Sarovar Dam in Gujarat, their legitimate claims to resettlement and rehabilitation having been officially ignored (Bavadam 2003). Thus, those on the edge of subsistence are disproportionately represented in the figures of those suffering from every kind of water-related calamity, including reservoir submergence, cyclone, and flood damage. Typical strategies adopted by marginal farmers in the face of disaster include seasonal migration from the countryside to the towns in search of work, reduced food consumption, and the sale of whatever assets they possess—jewellery, keepsakes, timber or other natural products. In western Orissa, where there has been a marked recent increase in the number and severity of droughts, district collectors have reported that families respond by removing their children from school, resort to child and bonded labour, and experience strain in family relationships (Bandyopadhyay and Das 2002).

Among drought victims in Rajasthan, powerlessness and exclusion are so extreme among certain tribal populations that people do not enrol for the rations and food-for-work programmes to which they are entitled (Stuart 2003). These programmes are also not as adequate as they should be. In early 2003, as the dry season took hold, only a fraction of the resources sought by state governments from the National Calamity Contingency Fund had been released (Kang 2003). As some commentators, notably P. Sainath, have ruefully observed, the benefits of drought-relief programmes do not always reach the end of the road where the truly indigent are to be found (Sainath 1996). The plight of such marginalized families, always dire, typically near the edge of survival, is rarely on the radar screen of mainstream popular attention. The stories that do appear in the media usually make little lasting impression. They belong to the bad old days of which India does not wish to be reminded, when the country could not manage to feed itself. Those images have been banished to the pages of history, despite the reality that many people suffering from natural resource-based calamity or dislocation due to 'development' fall-out, continue to have difficulty feeding their children.

For some observers, this is an unfortunate but necessary corollary of the economic process which upgrades lifestyles and incomes for a large proportion of the population. The losses of land, water, and livelihood endured by a small proportion of the population were lamented, but endorsed, by Indian leaders such as Jawarharlal Nehru and Indira Gandhi, as being in the greater national interest (Mander 2003). For others, social justice requires that 'development' should not further divide the gap between gainers and losers. An end to penury among the least disadvantaged, especially now that state-provided safety nets are being disbanded to make way for the efficiencies of market forces, is, in this view, the very thing that 'development' is supposed to have ended.

Although controversy surrounds almost every aspect of India's deepening water crisis, from the analysis of causes to the advocacy of prescriptions, there is clear unanimity that such a crisis exists. The evidence of groundwater over-exploitation, and of chemical and bacteriological contamination are causes of widespread alarm. The growing frequency of localized conflict over water, in the form of protests against the construction of major dams at Tehri and along the Narmada river, of angry protests and popular violence over reservoir or pipeline releases in the Kaveri basin, of widespread water riots at the height of the dry season, and of popular activism against the commercial ownership and exploitation of water, have become part of the context in which policies relating to livelihoods and lifestyles are debated. At one end of the spectrum are the environmental campaigners, who believe that modern development—commercial forestry, mining, monoculture agriculture, intensive use of fertilizers and pesticides, atmospheric pollution—have abused the earth and destroyed its capacity to receive, absorb, and store water (Shiva 2002). They perceive the loss of time-honoured methods of rainwater harvesting and storage as the crux of the problem, and believe that the resuscitation of non-energized irrigation techniques, multi-cropping with drought-resistant millet and sorghum, and other traditional and ecologically-sound resource use and management practices are the ingredients of a more sustainable approach. In this ideological view, natural processes and techniques, and the traditional social and organizational structures with which 'development' has interfered, need to be recovered.

At the opposite end of the spectrum lies the belief that the application of technology and large-scale investment in developing and exploiting the resource base, along with the manipulation of supply and demand through the involvement of commercial forces, are the best way forward. The proponents of major infrastructure development—mega-dams, canal networks, and pumping stations of a scale never before seen anywhere in the world to transfer water between major river basins—believe that their schemes can be adjusted to respect environmental integrity and cushion or compensate social distress. These are seen as practical problems of implementation which can be solved, even though the record with existing large projects has not been exemplary up to now to say the least. Of the 25 million people displaced by development projects since 1950, less than 50 per cent have been rehabilitated; the rest were pauperized by the process (Jain 2001). However, such realities are often seen merely as uncomfortable corollaries of an unstoppable process of positive advance. In today's dominant ideological view, the vagaries of nature with which India's poverty has down the centuries been so intricately allied are there to be tamed and defeated by macro-technological application. Ancient ways of interacting with them are to be superseded, not recovered.

Many exponents of human development—and Unicef is an obvious organizational example—believe that micro- and macro- approaches are not contradictory. Both have their merits. The difficulty is how to participate in the many debates in such a way as to remain within a harmonious ideological spectrum. This is likely to become more problematic in India as water stress increases. What the international exponents of water as an economic resource with a price tag failed adequately to realize when they first set up their stall in the early 1990s is that, in all settings with severe resource constraints, political controversies are bound to develop from the bottom to the top of society among the interested parties. These cannot be confined to differences of economic, scientific, or technical point of view. Whether the issue is land submergence, water privatization, hiking of rates for water or electricity tariffs to pump it, the viability of massive engineering projects, the impact of global warming on rainfall and temperature patterns, or the local ownership of water and water installations, the debate inevitably takes on a political and ideological cast. As new contests over this ever more pressurized natural resource develop, differing viewpoints may well draw further apart, and the forces and constituencies they represent take up ever more confrontational positions on a 'water battleground'.

It is quite possible that the future management of water resources will become an arena in which the exponents of small-scale, locally-adapted participatory approaches, and advocates of large-scale, high-tech, economically magisterial types of intervention, end up on a damaging collision course. The environmentalist NGOs, popular movements, and community groups can be expected to adhere to a people-centred, local ownership, socially equitable version of water resources management, based on better capture and conservation of precious rainfall as close as possible to where it lands. This can be on the earth, or on any intervening structure such as a rooftop, from which it can either be collected and stored or used to recharge aquifers in the immediate neighbourhood. Most of the institutions of central and state government, international banks, and external support agencies are more likely to align with the larger, more instrumentalist view, in which billions of cubic metres of water flow are manipulated and re-allocated courtesy of large infrastructural projects. In this scenario, state institutions and commercial corporations will increasingly claim ownership of water and play leading roles in its distribution process. In spite of rhetoric to the contrary, community responsibilities towards water resources and their management will not be taken seriously—except when it comes to paying the bill. This they will increasingly be expected to do, whatever the quality of the service.

However, policy choices are rarely clear-cut when so much is at stake and there are so many not only interested, but desperate and driven, parties. In such a vast, increasingly consumer-conscious and essentially democratic country, many diverse—even contradictory—policy streams will inevitably be followed in tandem. Many urban municipalities stretching the length and breadth of the country from Aizwal in Mizoram, to Chennai in Tamil Nadu, are now actively

promoting rainwater collection on rooftops and other water harvesting ventures. Water bodies within cities which used to be filled in and
built over are now being restored and protected, often under court
orders. Digvijay Singh, the then chief minister of Madhya Pradesh,
had taken up the water harvesting baton with serious political commitment. As a first step he instituted a state-led programme to regen-
erate the chronically drought-prone tribal district of Jhabua by
mini-watershed management, which in six years nursed a 'moon-
scape' back to life (Agarwal et al. 2001). In Karnataka's Kolar district,
state government water officials identified disused traditional tanks
by the use of satellite imagery. By de-silting the tanks and their feeder
channels, water was restored to previously dry wells and the area's
agriculture reinvigorated (Luce 2003). Small- and medium-scale
ventures of this kind are now legion throughout the country, with the
support of not only NGOs and environmental campaigners but also
of government officials. They are backed by some distinguished
senior policy advisors, ex-members of the Planning Commission,
senior politicians, and departmental secretaries.

There are signs, therefore, of increasing political and official commitment to reducing the flow of India's rain into the sea by capturing
it where it falls, as a more cost-effective method of increasing utilizable water resources than major construction projects or uncontrolled
water-well drilling and pumping. However, it is also undeniably the
case that many officials only perceive water harvesting as of marginal
benefit in the grand scheme of things. Persuading the water establishment in India to regard this as a serious developmental option, whose
cumulative effect has implications for the whole water resources
picture, is difficult (Iyer 2001). No network of canals constructed as
a result of river-linking could ever manage to bring water within reach
of more than a proportion of farmers in drought-prone areas (Parsai
2003). As L. C. Jain, a former member of the Planning Commission has
observed, two-thirds of the terrain would remain uncovered and
would have to be augmented by local water harvesting programmes.
But ingrained assumptions about the importance of major inter-
basin transfers and their corollary, massive engineering works, for
resource redistribution between water-short and water-abundant
parts of the country, are hard to erode. Simple, low-cost, community-
led water harvesting initiatives are not taken seriously enough within

the establishment, not least because the approach appears to contradict the ideal of technological mastery and industrial might which has been a source of inspiration for so much of India's development up to now. It is, unfortunately, the case that such small-scale, low-cost constructions have far less appeal than large-scale endeavours in the contractor-oriented climate of many development undertakings.

As concerns about water's management and distribution have widened, and the survival, livelihood, and environmental implications of water stress and scarcity have grown, Unicef has tended, with some notable exceptions, to stay aloof from the major debates surrounding water as a resource. This area of its programme assistance and partnership with government and other institutions was, for Unicef, always to do with public health and reduction of child sickness. The public health dimensions of inadequate water supplies have not disappeared—indeed, in the context of chemical contamination, they have dramatically worsened. However, outside the high fluoride and arsenic-affected areas, public health issues do not have the same urgency as the threat to their entire way of life confronting communities whose water sources and natural resource base are close to collapse.

It has proved impossible for an organization such as Unicef, concerned about the most fundamental questions of life, livelihood, health, personal development, and future prospects for children, to stand aside from the water crisis facing India. Water poverty is a definition of poverty itself. In fact, water poverty should be used as an indicator of socio-economic disadvantage instead of merely as a symptom of such disadvantage, as is now normally the case. This is borne out by the evidence that families reduced to penury by persistent drought reduce their food consumption, withdraw their children from school, contract them into bonded labour or early marriage, and enter a migratory lifestyle in which the fabric of family life disintegrates. Many women and children are often reduced to an itinerant existence providing dirt-cheap labour on the very construction sites and engineering works which produce reservoirs, canals, electricity, roads, and benefits for others and for the state. Unicef can hardly claim to be an advocate for children's and women's rights and not address the marginalization that water-related calamity, natural and man-made, has imposed on many rural families.

Since Unicef withdrew its general support for water-well drilling in the early 1990s, it has been revising its role in the context of water supplies for disadvantaged populations or 'problem villages'. Its most useful contribution has been to develop models for participatory management of water services at the community level, helping, via partnerships with local NGOs, to create viable management systems and enable communities to develop and run their own services to the maximum benefit for women and children. However, the aim of most of these programmes is still focused primarily on health, with disease reduction taking first priority. A secondary justification is to enhance the living and learning environment for the young child, at home and in school. But a commitment to the livelihood dimensions of household and community water security has yet to emerge as a major preoccupation of its programme of development assistance. Ironically, many of the most effective water-related activities in which Unicef has recently been involved are part of what brought it into water supply in India in the first place—emergency drought relief.

During the mid-1990s, Unicef began to bring the vocabulary of 'environment' into its programmatic concerns surrounding water supply and sanitation. In fact, it renamed its water and sanitation programme 'Child's Environment'. This change sounded as if Unicef was taking up in earnest the new emphasis on 'sustainable development' which emerged so vigorously in the aftermath of the 1992 Earth Summit. The 'environment' to which Unicef referred turned out to be the child's living, nurturing, and learning environment, not the protection of the natural resource base with which the term is normally associated.

However, in the light of the concern in India about the 'unsustainability' of handpump–boreholes and the growing threat to drinking water supplies from over-extraction of groundwater, there was also some Unicef engagement with the protection of the freshwater resource base. In 1995–6, research collaboration began with the WWF on 'Freshwater for India's Children' (see Chapter 7). This reflected an effort to develop new partnerships and expand Unicef's range of water-related concerns, in the international NGO world and with national bodies.

At the same time, changes in donor preoccupations moved Unicef in a similar direction. During the late 1990s, Unicef initiated a series of pilot schemes for small-scale water resources management with assistance from the Dutch government. Unicef's major bilateral donors—the Swedes, British, and Dutch—were all enthusiastic supporters for protection of water resources through micro-watershed management, artificial recharge, and community water harvesting. For a time, it appeared that the natural environment, at least in the context of rural water supplies, was going to become something of a Unicef preoccupation, with backing from environmentally-conscious donors and partnerships with Indian institutions and major NGOs such as the CSE.

The new Unicef-assisted programme was known as 'Environment Protection and Water Resources Management' and its four pilot projects were in Madhya Pradesh, Tamil Nadu, Gujarat, and Maharashtra. These were all states in which the over-exploitation of groundwater had lowered the water table and rendered many existing handpump–boreholes unusable. The idea was to try out watershed management techniques in different hydrogeological conditions in the hope that successful demonstration would encourage states to replicate the techniques on a wider scale. Integral to the proposed plans for all four pilot projects was the idea that communities and local NGOs would be involved, and that communities themselves would be inspired and assisted to take on the task of managing their own local freshwater resources (UNICEF 2001). The intended recharge structures—check-dams and percolation ponds—were not expensive to build, nor did they require elaborate O&M. With a minimum of local organization and training, community management would surely be straightforward. But, as so often occurs, there was a gap between the plans and what actually transpired on the ground.

Only in one state—Gujarat—was there any serious attempt to carry out the project in a genuinely participatory way. Here, the water table had dropped by 10–15 metres in some areas, and source depletion was so serious that 70 per cent of the state's population faced potential shortage. A high proportion of India Mark II and Mark III handpumps installed under the state programme had run dry and the Gujarat Water Supply and Sewerage Board (GWSSB) was having great difficulty in locating new sources of supply. The locations selected for the project were two chronically water-scarce talukas of Rajkot district situated in the Vanchiyawadi watershed. This had been already designated a grey zone for groundwater: 75 per cent of the resource had already been extracted. The Vanchiyawadi river and its tributaries seemed a good candidate for recharge since the river flowed for four to five months with a sufficient stream for harvesting with check-dam and dyke technology, and there was no pollution. Perhaps the chronic water shortage in the area was inspirational in helping the project towards success, but no less so was the quality of leadership. The GWSSB, the Gujarat Jalseva Training Institute, the engineering staff in Rajkot's public health department, NGOs, and Unicef's state office were all committed.

The whole process from its inception was conducted in a participatory way with the involvement of village leaders, panchayats, local-level functionaries, and health and anganwadi workers. Every effort was made to inform the villagers about groundwater depletion and motivate their involvement in the new construction measures and in the protection of existing sources. Meetings were held to engage community leaders and village representatives in site selection for check dams, dykes, and flow diversion channels, with the sarpanch of each village in charge of construction contracts. With the help of NGOs and motivated women leaders, existing sources were cleaned and improved, wells disinfected, and measures introduced to promote sanitation and hygienic use of water. In each project village, a Village Development Committee was constituted to manage and maintain the water-collecting installations, handpumps, dug wells, and other water-related services for the future. An energetic awareness campaign was conducted in each village soon after any new construction, led by the district public health engineering team, with the help of teachers, health staff, *kendras* (centres), and village-level workers.

In three of the four pilot projects, the recharge structures were effective from a technological point of view. Only in Madhya Pradesh, where the work had been poorly sited and implemented, was there no marked improvement in the water table level. In Gujarat, water table depths had improved by up to five metres in the post-monsoon season, and the number of months in which villagers had been dependent on tanker supplies because of dry wells and boreholes had gone down from six or seven to one or two. In Maharashtra, the average rise in water levels since the pre-project period was 4.70 metres. In the villages of Gujarat, there were many advantages for the villagers, not only in drinking supplies but in milk, fodder, and agricultural production. However, there was some negative feedback in Maharashtra that some of the *bandharas* (small check-dams) had been built where they only benefited a handful of better-off farmers. And even in Gujarat, where every village had established a Village Development Committee and all contractors used were local, there was little evidence of the desired sense of community ownership and responsibility for future O&M. As far as enabling water harvesting to take off was concerned, the objective was over-ambitious and needed a far longer and more systematic commitment. Unless local NGOs were prepared to take it further, or state or district PHEDs proved responsive to community requests, once the project was over the story was depressingly familiar—everything proceeded much as before.

Without sustained inputs of motivation, capacity building, and financial resources, it would require a lot more than a few pilot projects to effect a radical switch into a water resources management regime based on local watersheds. Only in Gujarat had there been any real attempt to try.

When Unicef had entered the rural drinking water scene in India, offering start-up resources and the introduction of a new technology, it had been pushing at an open door. There had been in the 1970s immense receptiveness to the handpump–borehole as the answer to drinking water problems, and within a short time the technology had been adopted, absorbed, and massive government resources put behind its universal spread. In the case of water harvesting in the 1990s, very different circumstances prevailed. There was no problem with the technology. On the contrary, check-dams and percolation wells were relatively inexpensive, and local knowledge combined with hydrogeological expertise and equipment could easily be deployed in every part of the country to choose sites and provide technical and management supervision. But the official and political willingness to commit to micro watershed management as a water resources development strategy for 'problem villages' remained elusive. There was no expanded definitional analysis of 'problem villages' to take account of community water resources, no targeted plans for coverage vis à vis rainwater harvesting and groundwater recharge.

When pilot projects to bring in a new technological and organizational approach do not lead to systematic take-up by the institutions of government, Unicef has difficulty knowing where to go next. Since those early ventures—early in terms of Unicef's involvement in watershed management—Unicef has continued to support similar studies and pilot schemes in other states. In Maharashtra, for example, where for the past four years 12,500 villages have faced chronic scarcity, Unicef has recently embarked on some new pilot and demonstration schemes (Groundwater Surveys and Development Agency 2003). But these have tended to be peripheral to its main 'child's environment' thrust with its emphasis on healthy living, learning, and nurture. In 1998 came the publication of the joint Unicef–WWF Report, *Fresh Water for India's Children and Nature*. This failed to achieve what its advocates had hoped for—to propel Unicef in India, and perhaps elsewhere, into systematic support for fresh water resource management at the community level. Some of those in the organization's 'child's environment' team who had become increasingly concerned about India's deepening water crisis were disappointed. They regretted that they had failed to capture Unicef's, or the government's, imagination for addressing the current water challenges facing India's poorest villagers in a thorough and strategic manner.

One reason was that the policy environment within the organization had radically changed. Unlike the early days of support for rural water supplies, when enthusiasts in Unicef India were able to persuade senior policy-makers in New York to make a leap of faith, this time they failed. The India programme had opened the way for Unicef to fill a niche in providing water to rural communities all over the world during the 1970s and early 1980s that no other UN or international organization was then attempting to fill. Now another pioneering leap and a change in international policy direction concerning water was required. In the early 1990s, Unicef's senior nutrition advisors had been able to make a conceptual shift in nutrition policy, moving from an exclusive preoccupation with child health, and adding a livelihood dimension by emphasizing the need for 'household food security'. However, the same shift for water, away from an exclusive emphasis on its role in disease reduction, towards the need for 'household water security' (quantity as well as quality) was not now achieved. There was no sufficiently powerful voice making the case at the international level, nor a receptive policy atmosphere for it.

In India, however, drought in its remorseless way provided the context in which household and community water security simply had to be addressed. In 1999, Gujarat once again suffered from poor and erratic rains, leading to one of the worst droughts it had ever experienced (Government of Gujarat 2000). All the reservoirs in Saurashtra and Kachchh ran dry, and in many regions groundwater was the only source of water for basic survival needs. Excessive pumping of groundwater continued, leading to further alarming drops in the water table. Around 25 million people were affected, and 9500 villages, four cities, and 79 towns faced acute water shortage for drinking, maintenance of hygiene, and for their livestock. There were huge losses in agriculture and of livelihood opportunities. Handpumps repeatedly failed. Women and children were forced to travel many kilometres to collect a household supply, or fodder for livestock. In spite of emergency relief programmes—drilling of boreholes, relief works, cattle camps—over 3000 villages were dependent on tanker supplies. This was a year in which water hardship led to conflict, and public unrest broke out between rural and urban populations in and around the cities of Rajkot and Jamnagar with large-scale movement of people and livestock in search of water and fodder, including from neighbouring Rajasthan.

The Gujarat government requested advice from a joint UN mission led by Gourishankar Ghosh. Ghosh was well known in Gujarat as a previous water secretary, had since that time been mission director of the Rajiv Gandhi National Drinking Water Mission, and was now Unicef's Chief of Water, Environment, and Sanitation. The mission was to look beyond immediate emergency relief to longer-term measures for drought mitigation, including the potential of rainwater harvesting and groundwater recharge. These techniques were becoming more common in different parts of the state, and some NGOs had become renowned for their water-related work. For example, the Sadguru Water and Development Foundation based in Dahod district had long been working with adivasi communities to promote sustainable development and use of water resources by means of lift irrigation (Kumar et al. 2000). They had conclusively demonstrated that local rivers which dried up soon after the monsoon could be regenerated into perennial streams by building check-dams along their lengths. This created a cascade of small reservoirs, and by holding back the water in the landscape, regenerated the year-round farming economy. This also ended seasonal out-migration in entire communities (Jagawat 2000). The mission visited Dohad and focused on all aspects of water assessment, management, and use. As a result of its preliminary findings, Unicef commissioned a 'White Paper' on the state of water resources and water-related services in the state.

The key conclusion of the Gujarat White Paper was that, since drought was a cyclical phenomenon, efforts should be targeted less towards 'drought mitigation', and more towards the longer-term solution of 'drought-proofing'. A thorough overhaul of policies was recommended, including more emphasis on the effective *management* of water resources, and less emphasis on the *development* of resources via large-scale projects, some of whose outputs were neither technically efficient nor cost-effective. The management systems proposed were the establishment of river basin organizations and, at lower rungs of the administrative ladder, of watershed committees, urban water councils, and water-user associations. Mass campaigns to promote awareness about water scarcity, its causes, and water conservation techniques were also recommended.

In the wake of this 'White Paper', Unicef prepared a proposal in March 2001 for short-term and long-term assistance for women and children in the drought-affected states of Gujarat, Madhya Pradesh, Orissa, and Rajasthan (UNICEF 2001). Short-term measures included pumps, pipes, tankers, well rejuvenation, water storage, and its transport; the promotion of hygiene; and relief employment on water conservation projects. Long-term measures included drought-proofing by water harvesting, recharge, and local management of drinking water sources by users; advocacy for groundwater regulation and conservation. Included in both short-term and long-term proposals were a range of measures to support women's groups, child nutrition, education on water-awareness, institution-building in communities, and communications activity. Since that time, Unicef's activities in these and other drought-affected states have included various kinds of support for drought-proofing and drought mitigation, linked, wherever possible, to ongoing programmes of support for 'child's environment' activities, including school sanitation.

Rajasthan was the state worst affected by the poor monsoon of 2002. In the state as a whole, the rains were deficient by over 50 per cent, with 17 out of 32 districts receiving less than 60 per cent of their normal expected rainfall. Already by September 2002, 50 per cent of handpumps in the state were dry; 27,000 villages were facing shortages, with hundreds needing new boreholes and existing open wells and boreholes deepened (UNDP 2002). Over 10,000 villages were already being supplied with water sent by train and truck, and this was at a stage when the dry season had hardly begun. One of the most severely affected districts was Tonk, lying just south of the capital, Jaipur. Here, within the context of its existing 'child's environment' programme, Unicef played a useful role in assisting communities to face an unfolding short-term and potentially long-term water shortage crisis.

Tonk is a critical or 'dark' zone for both water quality and water availability. In October 2002, as they faced a winter and spring with almost no water reserves, every community in this drought-stricken area had one refrain. They were happy to review their progress in environmental sanitation, explain the activities of their 'user groups', show women's ability to take handpumps apart, demonstrate the school's commitment to cleanliness and a hygienic way of life. But at the end of any presentation of community water and sanitation action, the local sarpanch always reverted to the same subject: harvesting water. All wanted to construct *anicuts*—water-retaining check-dams in the beds of seasonal rivers. The benefits of capturing the rain as close as possible to where it falls have become much more widely known since the TBS in neighbouring Alwar district began work over 15 years ago. Today, quite substantial earthen and concrete structures across the courses of ephemeral rivers, small low-lying bunds, and pipes to divert water from waterlogged fields directly into wells, are becoming familiar features of the landscape. Villages in stress have ceased to believe that any day soon a great canal will arrive from a neighbouring state to spill water into their fields. Instead they are keen to take the re-greening of their land into their own hands. The positive effects within one season of undertaking water retaining measures are well known even to people way off the beaten track.

Water conservation is also the theme of the District Collector, Ashwini Bhagat, who described this as the new direction for water policy. The government's drought-relief programme, which regularly employs people on public works such as road-building in return for food rations and other support, now includes anicuts in the range of community construction projects for which emergency drought employment can be offered. In late 2002, where social mobilization under the existing Unicef-assisted programme had already been undertaken by a local NGO, anicuts could be added to the community package under the budget heading of emergency employment works. According to Bhagat's own statement on 11 October 2002, over the past two years, he had sanctioned the construction of 70 water conservation schemes under the rubric of 'emergency relief', whereas the number of new water extraction schemes sanctioned was only eleven. 'Even in this extremely dry year, water harvesting has led to recharge of wells and aquifers, to the extent that farmers now have some water for irrigation. This has encouraged them, and the mood of the villages is buoyant.' He sees water harvesting not only as a benefit in producing more water, 'but also in creating a sense of empowerment among the villagers. They have recovered their traditional practices for water conservation, and accord them new respect.'

The perennial drought emergencies on small and large scales, and the crisis of groundwater depletion, are giving a new spur to Unicef to engage in drought-proofing and community water crisis preparedness and management. Working with NGOs in what has become the tried Unicef formula, undertaking participatory appraisals and local water source mapping exercises, assisting with rainwater collection, harvesting, and conservation, and establishing local water organizations to assume management responsibilities, are important components of drought-proofing (UNICEF 2002). People have begun to be persuaded that they are jeopardizing their own water security by taking more water out of the ground than the natural cycle of rainfall and run-off is capable of putting back in. Their growing understanding is essential to change.

Even more important is that the government in all its forms needs to support the policy of community water security that certain district collectors and other officials are keen to promote. Significant financial resources will have to be found, and the entire administrative apparatus reoriented towards effective, sustainable, and equitable water management. Something so vital for India's future should not be left to NGOs and other incidental players, and treated as marginal to the main water resources policy agenda. If the official mindset continues in 'business as usual' mode, India's freshwater crisis is more than likely to get worse.

There can be no more important priority for water-short communities than to find ways to reverse the deterioration in water sources that threatens their children's survival, health, and future livelihoods. Growing interstate and inter-community conflicts over water already looms. In the end, pre-empting and adjudicating such struggles, promoting water conservation, and allocating resources fairly between the better-off and the rest, is what the management of water for life and health will have to be about.

After 35 years of support for water and environmental sanitation in partnership with India's central and state governments, the water-well industry, and NGOs, Unicef's experience contains many pauses for reflection. The pattern of improvement in any area of human affairs is never seamless, and there have been many involuntary setbacks as well as solid achievements. As this book has demonstrated, it is part of the internal dynamics of any development programme that each advance down a new and promising path throws up new sets of challenges—technological, financial, managerial, socio-economic, cultural—which in their turn need to be addressed. Solutions to one set of problems—for example, the introduction of hard-rock drilling to meet emergency water shortage—may, over time, augment problems such as groundwater over-extraction and chemical contamination which need to be addressed in the next generation. Some initiatives with the potential to 'go to scale'—CDD, school sanitation, fluoride filtration, or rainwater harvesting—may become stymied because commitment is not forthcoming from the government or the commercial world to back successful models in a timely fashion with funds and energetic promotion.

These dynamics would operate even if the parameters within which programmes function remained static—which they do not. All the time—and in India from Five-Year Plan to Five-Year Plan and from election to election—the political, economic, budgetary, social, and environmental context in which programmes are implemented is changing. There are also evolutions, nationally and internationally, in knowledge, understanding, and analysis surrounding the 'problem', along with fashionable new doctrines and practices proposed for the 'solution'. In the past 20 years, there have been many important innovations in international thinking regarding water and sanitation, not least the switch from a sense of complacency about a freely available resource to a concern about increasingly scarce and degraded supplies, and the first murmurings about the threat of future 'water wars'.

The idea of water as a precious resource which should carry a price tag, the need to move away from supply-driven to demand-driven service design, the importance of effective 'software' as well as 'hardware' within programme delivery, and the need for integrated water resources management at all levels—from the community, to the state, to the nation; all these ideas were embryonic at the beginning of the 1990s, but now occupy a key place in water-policy thinking. Another important change is that sanitation and hygiene are no longer seen as the poor relations of 'water supply'. Instead, they are recognized as essential to human dignity and health. Even between the second World Water Forum, held in 2000 in the Hague, and the third held in 2003, in Kyoto, ideas had changed. Instead of compartmentalizing 'water for health', 'water for food', and 'water for the environment', a more holistic goal is now in sight: water security and sanitation for all, especially the poor.

The Unicef-assisted programme in India, as well as reflecting many of the changes in thinking which have characterized the world of water during the recent past, also demonstrates that mindsets and circumstances thought to be unassailable can—and do—change. The Medinipur experience may as yet be exclusive to West Bengal. But it has demonstrated conclusively that there can be a strong desire, manifesting itself as a market even among the most cash-starved rural families, for home toilets. It took determination to break down entrenched attitudes, and painstaking work to build local organization, to make this happen. But it has happened. And if it can come about in one setting, with a suitably adapted approach, resistance in another may similarly be broken down. Moreover, the conviction that women could be trained to fulfil managerial roles in community water services has shattered entrenched gender-based prejudice. In villages throughout the country, women take their place on water and sanitation committees. They also dismantle and repair local handpumps, working in teams alongside men—a prospect which few would have considered possible a generation ago. Today, scepticism greets the idea of ecological sanitation because the principles behind the technology run counter to ingrained cultural practices regarding human wastes. But within a few years, at least in congested and high water table areas, it is more than likely that this approach—with its health, anti-pollution, and livelihood benefits—will start to catch on.

It is important to recognize, as is clearly shown by the story of Unicef assistance to rural water supply, that any solution proposed in its time as virtually universal—such as the handpump–borehole for safe drinking water in hard rock areas—has its limitations. However successful a 'mass' supply-driven programme appears to be, it should not be promoted without reference to the many social, economic, and environmental variables belonging to different settings: from village to village and block to block, let alone from district to district or state to state. The hydrogeological parameters for which it seems an ideal response may be more or less identical, but as experience has shown they should not be allowed to drive out 'softer' considerations. In the end, a water supply service consisting of stand-alone installations rather than a network of pipelines operated by a central command will have to be operated and owned by the local community, albeit with supervision and technical support from outside. Whether the nature of these installations and the service they offer reflect a community's desires will ultimately govern motivation to manage them in such a way as to distribute benefits and responsibilities suitably, and keep them in repair.

Declining water tables and drying boreholes are one reason for a high proportion of abandoned handpumps. Technical breakdown, slow response, and belief that the government should repair 'its' installations are others. But these are not the full explanation. In too many cases, handpump–boreholes have been provided to communities without reference to their own sense of requirements, the state of their existing water sources or preferences, and without any discussion or calculation as to whether maintaining this relatively sophisticated technology—compared, for example, to a dug well, check-dam, or tank—is within their financial means.

The moment a piece of machinery is erected between a community and what it has previously exploited as a free natural resource, the whole economic and political picture surrounding that resource alters. The machinery, whether it is a handpump, a power pump, a flour mill, or a vehicle, has to be owned by someone and managed by someone. Moreover, its maintenance costs money and needs technical knowledge and skills. Spare parts such as washers for handpumps and other minor items such as lubricating oil are within the reach of most rural communities, even the poor; but major repairs are another matter. And there may be disputes surrounding this equipment: whose land is it on, who gets to use it and at what times of day? In such cases where levies to pay for a service are exacted, the people who pay need to monitor that this money is indeed spent on making their services work, and does not simply vanish into some official or not-so-official pocket to be spent on something else. For a number of reasons, there may be reluctance to assume responsibility, including financial responsibility, for something which, in the absence of prior organizational arrangements to deal with such issues, can end up by being as much a liability as an asset.

This goes to the heart of the matter: household and community 'demand'. In many communities where handpump and borehole installations are not much used, it is often because people have other sources of drinking water which they prefer. Their failure to report breakdowns is not always the product of passivity or ignorance. It is a simple statement of consumer choice. If this were not the case, why is it that an estimated on-going failure rate of 30 per cent—or higher—for handpumps does not produce a devastating drinking water crisis in areas other than drought-stricken ones? The idea that virtually every handpump–borehole provided under the central- and state-sponsored programme has, over the years, met, or had the potential to meet, child health and survival needs makes excellent publicity for donor organizations and politicians alike. But it is far from being a true reflection of reality. And this is a verdict on official and donor unwillingness to look at livelihood needs, instead of focusing on the role of water in disease reduction and public health. More understanding of disease causation and of the virtues of 'safe' water would change attitudes towards the value of handpump–boreholes in some locations.

The failure so far to make a genuine transition to 'demand-responsive' service structures in many of the areas now under sector reform needs urgently to be rectified. There is a continued reluctance in many parts of the rural drinking water establishment and within the bureaucracy to perceive villages as 'consumers' of services instead of merely as their 'beneficiaries'. If villagers are to be expected to shoulder service costs, the services they receive will have to correspond to their social, economic, and livelihood requirements. In all locations where a borehole is to be drilled and a handpump installed, the demand for the service, and the local capacity to maintain it—skills, organization, and financial resources—need to be fully established before the drilling rig is ordered, let alone arrives. The same applies in the case of all types of community facilities and installations, including check-dams, sanitary marts or production centres, and school water and toilet blocks.

The long experience in India with 'vertical' programmes and the 'supply-driven' service model which viewed the poor as humble recipients of state-generated largesse is very difficult to reverse. But if there is one lesson that has been learnt from Unicef's 35 years of water and sanitation programming in India, it is that without real involvement and commitment from the people, especially the women, no water or sanitation service can be sustained. As time goes on, the same will inevitably be the case with the resource itself. Without the involvement of its consumers in conservation, protection and allocation, the freshwater resource will not be sustained either.

So much lip service in official documentation and international discourse has been paid to the need for community management and ownership in water and sanitation services that this seems a mundane note on which to end. But it is far from being a mundane issue. It is ultimately the critical issue. As experience shows, any programme which attempts to apportion water or water services on the ground, and in which payments such as fees or water rates are involved, requires minute political negotiation and a day-to-day local capacity to deal with water releases, pipeline flows, irrigation take-up, rules and regulations, fines and punishments for infringement. Service installation can be delayed for months by local disputes about the siting of drains for storm water run-off, community stand-pipes, and the amounts to be levied from different members of the community. But if such decisions are hurried or pre-empted from outside the community, the programme may fail. All such delays run counter to preoccupations with time-bound targets, machinery deployment schedules, construction timetables, and other bureaucratic and cost-efficiency considerations.

There are trade-offs, too, concerning quality. If the toilet pan and trap, or the pump and its installation, collapses or breaks down within a few weeks, the service is discredited. When local PRIs are responsible for hiring contractors and purchasing equipment, they may choose to cut corners which ultimately provide them with a less-than-ideal service. But as consumers, these are things they have to learn. And in the end, they are more likely to make decisions that best serve the interests of their constituents. When they make mistakes or find they have been cheated, they will learn how to resort to the courts and other institutions of redress to defend their own investments. That is what a politically aware, democratically constituted, consumer society is about.

The intricacies of community management and ownership are complex, and it can take painstaking effort to develop mechanisms on the ground that genuinely enable women, children, households, and communities to solve the water supply, cleanliness, and waste disposal problems they face. But in spite of all the difficulties, it has been Unicef's experience to find within the myriad experiences of drought-prone and flood-prone communities much cause for optimism. Where the energies and enthusiasm of local panchayati leaders, women's group members, schoolteachers, anganwadi workers, handpump mistris, water user groups, block development officers, health care workers, working farmers and their families have been harnessed, it is heart-warming to witness what they can achieve. Even while schemes operating at the macro-level provoke huge controversy, public opposition, mass activism, and violent dispute, at the micro-level where there is often nobody grand to notice, many extraordinary accomplishments are underway. At their best, the technical and organizational support from local NGOs, the backing from block and district officials, and the funds and occasional advice from Unicef, seem almost incidental to the process. The real 'development' that has taken place is that ordinary people, many of them poor or very poor, have managed to gain control over a key environmental resource, and manage and use it in ways which give them more security over their lives, and the health of themselves and their children.

As India's water crisis becomes increasingly susceptible to clashes of interest and ideology in national debate, increasingly difficult questions—about maintaining social stability, defusing conflict, ownership, rights and responsibilities, the appropriate devolution of power and financial say-so—will inevitably be asked. Ultimately, the question on the bottom line will turn out to be: Who owns the rain? The answer is that those who are today making the most of the rain where it falls—on the ground, in the soil, under the ground, and in the streams that pass by their homes—must be allowed to 'own the rain' on which their health and livelihoods depend.

Advocacy for community management and ownership of this vital natural resource, and of services based upon it, is essentially advocacy of a political and democratizing process. This is an inescapable reality. But it is not an easy reality for an organization such as Unicef. As a humanitarian organization focused on children and their rights, Unicef holds hard to the principle of staying above all political controversies, including those concerning access to, and ownership of, natural resources. But the potential for confrontation between those who have traditional rights in the resource but no economic or political clout, and those who have plenty of clout and little respect for ancient value systems, nonetheless threatens.

Out in the byways of Tonk, Howrah, Mysore, Betul, Rangareddy, Erode, Dhule, Medinipur, and many other towns and districts, examples can be found of communities where such issues are being resolved. Whether or not the word 'democracy' or 'equity' is applied, that is what community management and ownership are really about. Where everyone has a role in decision-making, is prepared to respect and negotiate differences, and the rights of the weak are given equal weight with those of the strong, a united front for clean, healthy, and productive living can be constructed. Only when that front spreads and widens will it be possible to attain the goal of access to adequate water and environmental sanitation even for India's most disadvantaged. After 35 years, that goal may still seem elusive. But the effort to drive towards it is infinitely worthwhile.

References

Agarwal, Anil, Sunita Narain, Indira Khurana (eds) (2001), *Making Water Everybody's Business: Practice and Policy of Water Harvesting*, CSE, New Delhi.

Athavale, R. N. (2003), *Water Harvesting and Sustainable Supply in India*, Centre for Environment Education, Ahmedabad.

Bandyopadhyay, S. K., and B. D. Das (2002), *Decentralised Planning for Drought Proofing and Sustainable Livelihoods*, Government of Orissa and UNDP.

Bavadam, Lyla (2003), 'Sardar Sarovar Dam: Woes of the displaced', *Frontline*, 12 September.

de Villiers, Marq (1999), *Water Wars: Is the World's Water Running Out?* Weidenfeld and Nicolson, London.

Dinesh Kumar, M., Vishwa Ballabh, Rakesh Pandy, and Jayesh Talati (2000), *Sustainable Development and Use of Water Resources: Sadguru's Macro-initiatives in Local Water Harnessing and Management*, Institute of Rural Management, Anand, July.

Government of Gujarat (2000), *White Paper on Water in Gujarat*, prepared for the Department of Narmada, Water Resource and Water Supply Department, Gujarat, by the Institute of Rural Management, Anand, with support from UNICEF.

Government of India (2002), *Water Supply and Sanitation: A WHO–UNICEF Sponsored Study*, Planning Commission, GOI, November.

Groundwater Surveys and Development Agency (2003), *Drought Situation in Maharashtra* (report submitted to Rupert Talbot), Pune.

Iyer, Ramaswamy (2001), 'Wanted: Fresh Ideas', in Agarwal, Narain, and Khurana (eds) *Making Water Everybody's Business*.

Jagawat, Harnath (2000), 'Revival Through Rivers', in Agarwal, Narain, and Khurana (eds) *Making Water Everybody's Business*.

Jain, L. C. (2001), *Dam vs Drinking Water*, Parisar Pune, citing Mid-Term Appraisal 2000, Planning Commission, GOI.

Kang, Bhavdeep (2003), 'Drought', *Outlook*, 5 May.

Lal, Murari (2002), *Possible Impacts of Global Climate Change on Water Availability in India*, Indian Institute of Technology, New Delhi.

Luce, Edward (2003), 'India plans river network to tackle floods and droughts', *Financial Times*, February.

Mander, Harsh (2003), 'A people savaged and drowned', *Frontline*, 12 April.

Murthy, Sachidananda (2000), 'Caught in a dry spell', *The Week*, 7 May.

ORG Centre for Social Research (2001), *Evaluation of Environment Protection and Water Resources Management Project in Gujarat, Maharashtra, Madhya Pradesh and Tamil Nadu*, final report, prepared for UNICEF New Delhi, July.

Parsai, Gargi (2003), 'River Grid Project will lead to New Inter-State Disputes', *The Hindu*, 17 May.

Pearce, Fred (2003), 'Conflict looms over India's colossal river plan', *New Scientist*, February.

Sainath, P. (1996), *Everyone Loves a Good Drought: Stories from India's Poorest Districts*, Penguin India.

Shah, Ashvin (1993), *Water for Gujarat: An alternative*, Sama Parivartana Samudaya, Federation of Voluntary Organisations for Rural Development in Karnataka.

Shiva, Vandana (2002) *Water Wars: Privatization, Pollution and Profit*, Pluto Press, London.

Stuart, Liz (2003), 'Liberalisation makes Rajasthan's drought lethal', *The Guardian*, 5 February.

UNDP (2002), *Situation Report: Rajasthan Drought*, UNDP, 12 September.

UNICEF (2001), *Drought in India: Challenge for Sustainable Development*, UNICEF New Delhi, March.

———— (2002), *Proposal for Drought Mitigation (Immediate Relief and Medium-term Interventions) 2002–4*, UNICEF Rajasthan; and discussion with Vinod Menon, UNICEF disaster Preparedness Advisor.

Glossary

Activated Alumina	Dehydrated, aluminium oxide used in filters to remove arsenic and fluoride from drinking water.
Adivasi	Indigenous tribal people who traditionally live in forested areas and depend on the natural environment for most of their subsistence.
Anganwadi	A crèche, or community childcare centre for pre-school children.
Anicut	A small dam built across a stream to impound water and provide an additional water source during the dry season.
Aquifer	Water-bearing geologic formation.
Arsenic	An inorganic toxic element, naturally occurring in certain geologic formations.
Arsenicosis	Disease caused by the ingestion of excess quantities of arsenic in drinking water. Prolonged consumption can lead to keratosis, skin pigmentation cancers and death.
Ashram	Hindu meditation centre.
Ayurveda	Traditional health care and healing science based on the use of medicines derived from herbs and plants.
Bandhara	Local term used to describe check dam in the state of Maharashtra.
Borehole/Borewell	Hole drilled in the ground for the abstraction of water by hand- or power pump.

Bund	Earthen embankment retaining a water body in, for example, a tank.
Check Dam	Small barrier constructed across a watercourse to restrict rapid run-off and to increase recharge to groundwater, especially in drought-prone, hard rock areas.
Chulha	Locally made cooking stove made of mud, whose fuel consists of wood, agricultural waste, and cow dung.
Collector	Senior government administrator of a district, sometimes also the district magistrate.
Dalit	Member of the scheduled caste, recognized in the Indian constitution as socially disadvantaged.
Down-the-hole-hammer	Air-operated percussive, hard rock drilling tool that travels down the borehole driving the drill bit to which it is attached.
Dracunculiasis	Commonly referred to as guinea worm, a disease caused by the ingestion of the water cyclops, *dracunculus medinensis* in contaminated drinking water.
Drill Bit	In the context of down-the-hole-hammer drilling, a hardened steel body with tungsten carbide inserts at the cutting edge. The drill bit fits into and is retained by the down-the-hole-hammer which drives it in to the rock face.
Drill Rig	Mechanically, pneumatically, or hydraulically operated machine for drilling boreholes. Drill rigs are normally truck-mounted for water-well drilling.
Fluoride	An inorganic element, naturally occurring in the geological formation of 19 Indian states.
Fluorosis	A crippling disease caused by the prolonged ingestion of excess fluoride in drinking water. Fluorosis manifests itself initially through mottled teeth in young children. In

	the later stages irreversible skeletal deformation occurs.
Fossil Water	Millennia-old water occurring in geological formations at great depth.
Ghat	An escarpment, hilly tract, or mountain road. In the context of lakes and rivers, steps constructed on their embankments leading down to the water level for washing and bathing. Ghats for cremation are also constructed in this fashion.
Gram Panchayat	Elected village council.
Gram Sabha	General assembly of a village.
Gram Samsad	Local council.
Gram Sevak	Village extension worker.
Groundwater Recharge	Replenishment of groundwater through the infiltration of rain water.
Groundwater	Water occurring in the geological formation below the water table.
Gurdwara	Place of worship for the followers of Sikhism.
Handpump Rejuvenation	Replacing a defunct handpump with a newer model.
Handpump	Hand-operated device for lifting water from borehole or other water source to a higher elevation.
Handpump–Borewell / Handpump–Borehole	Used synonymously. Handpump fitted to a borewell or borehole.
Hard Rock	Crystalline rocks of igneous and metamorphic origin such as granite, gneiss, quartzite, and basalt, with little or no primary porosity.
Hydrofracturing	A technique for revitalizing boreholes in hard rock that involves the injection of water under very high pressure to a low-yielding or dry borehole so as to flush out existing fractures or create new fractures to stimulate inflow to the well and improve yield.
Johad	Local name for small earthen dam in Rajasthan.

Kala Jatha	Street play performed by travelling theatre groups.
Kendra	Centre for information and coordination.
Kund	Below ground reservoir constructed to harvest, collect, and store rainwater. Unique to some desert areas of Rajasthan.
Mistri	Mechanic or mason.
Nadi	Open water body or pond in remote and desert areas of Rajasthan.
Naru	Local name for guinea worm in Rajasthan.
Overburden	Common term for regolith, the soil and weathered formation overlying bedrock.
Paan	Green betel leaf wrapped around betel nut and spices. Chewed as a palliative and digestive.
Panchayati Samiti	Assembly of Panchayats.
Panchayati	Administrative and development body of elected representatives.
Panchayati Raj	Local self-government system.
Pani Panchayat	Water committee.
Pukka	Quality product, well-finished. Strong, properly designed and built.
Rajputs	Warrior caste of Rajasthan.
Sankaracharya	Religious head of Hindu temple complex.
Sarpanch	Elected head of the Gram Panchayat.
Swajal	Name of a World Bank funded water and sanitation project in Uttar Pradesh that emphasizes community ownership, literally meaning 'own water'.
Swajaldara	Name of the Government of India's accelerated sector reform programme that builds on the Swajal project experience of community ownership, literally meaning 'own water stream'.
Taluk / Taluka	Part of a district, also called a 'Block', the smallest government-staffed administrative unit in India.
Tank	Traditional rainwater harvesting structure

	using natural slope of the land and with earth-retaining wall.
Tara Handpump	Direct-action shallow well handpump.
Temephos	A liquid chemical disinfectant sprayed in open wells and ponds to kill guinea worm cyclops.
Water Table	The depth below which the geological formation in the ground is saturated with water. The level at which water stands in an open, dug well.
Well Rejuvenation	A technique for flushing out an existing borehole with high pressure water through hydrofracturing or through the injection of compressed air.
Yatra	Journey undertaken for a religious or social cause.
Zilla Panchayat	Administrative and development body of local self-government at the district level.
Zilla Parishad	A council of elected representatives or a body of local self-government at the district level.

Index